VOLUME I

UV & EB Curing Technology & Equipment

VOLUME I
UV & EB Curing Technology & Equipment

R. Mehnert, A. Pincus, I. Janorsky, R. Stowe & A. Berejka

JOHN WILEY & SONS
CHICHESTER ● NEW YORK ● WEINHEIM ● BRISBANE ● TORONTO ● SINGAPORE

Published in association with

SITA TECHNOLOGY LIMITED
LONDON, UK

Copyright © 1998 SITA Technology Ltd,
Nelson House,
London SW19 7PA

Published in 1998 by
John Wiley & Sons Ltd
in association with SITA Technology Ltd.

All Rights Reserved.

No part of this publication may be reproduced by any means,
or transmitted, or translated into a machine language
without the written permission of the publisher.

Wiley Editorial Offices
John Wiley & Sons Ltd, Baffins Lane,
Chichester, West Sussex PO19 1UD, England

John Wiley & Sons Ltd, Inc., 605 Third Avenue,
New York, NY 10158-0012, USA

VCH Verlagsgesellschaft mvH, Pappelallee 3,
D. 69469, Weinheim, Germany

Jacaranda Wiley Ltd, G.P.O Box 859, Brisbane
Queensland 4001, Australia

John Wiley & Sons (SEA) Pte Ltd, 37 Jalan Pemimpin 05-04,
Block B, Union Industrial Building, Singapore 2057

A catalogue record for this book is available from the British Library

ISBN 0471 978906

CHAPTER I

RADIATION CURING: DEFINITION AND BASIC CHARACTERISTICS

I. INTRODUCTION
1. Definition of the term radiation curing 1

II. CHEMICAL SYSTEMS USED FOR RADIATION CURING 2
1. Acrylate and methacrylate systems 7
2. Cationic systems 8
3. Maleate/vinyl ether systems 10
4. Donor acceptor complexes 11
5. Unsaturated Polyesters 11
6. Thiol-ene systems

III. INITIATION OF CURING : ACTION OF PHOTONS AND ELECTRONS
1. Physical characteristics of EB and UV sources 12
2. Absorption of electrons and photons in matter 13
3. Initiation of UV curing by free radicals 15
4. Initiation of UV curing by cations 19
5. Initiation of UV hybrid curing 21
6. Initiation of electron beam curing: Free radical polymerisation 22
7. Initiation of electron beam curing: Cationic polymerisation 24

IV. UV (EB) CURING IN COMPARISON TO THERMAL CURING AND CONVENTIONAL DRYING 24

V. ECONOMIC AND ECOLOGICAL FACTORS FOR THE GROWTH OF RADIATION CURING 27

VI. REFERENCES 29

CHAPTER II

INDUSTRIAL APPLICATIONS OF RADIATION CURING

I.	INTRODUCTION	31
II.	RADIATION CURABLE COATINGS	33
	1. Applications on paper and paperboard	
	2. Applications on wood	
	3. Applications on plastics	
	4. Applications on metal	
	5. Applications on glass and ceramic	
	6. Miscellaneous coating applications	
III.	RADIATION CURABLE INKS	39
IV.	RADIATION CURABLE ADHESIVES	41
V.	RADIATION CURING FOR MANUFACTURING PLASTIC PARTS	42

CHAPTER III

UV CURING EQUIPMENT - POLYCHROMATIC UV LAMPS

I. POLYCHROMATIC UV RADIATION FOR CURING — 43

II. LIGHT EMISSION FROM A DENSE MERCURY VAPOUR PLASMA — 44
1. Light emission from a mercury gas discharge
2. Light emission from a microwave excited discharge

III. POLYCHROMATIC LIGHT SOURCES FOR UV CURING — 50
1. Low pressure mercury lamp
2. Medium pressure mercury arc lamp
   (i) Construction
   (ii) Spectral output
   (iii) Reflectors and lamp cooling
   (iv) Power supply and lamp control
   (v) Ozone-free lamps
   (vi) Lamp life
3. Microwave-powered medium pressure mercury lamps: Fusion UV Systems F 300, F450 and F600 — 71
   (i) Construction of the irradiator
   (ii) Spectral output
   (iii) Power supply
   (iv) Electrodeless vs. arc lamp: comparison of technical characteristics
   (v) Fusion UV Systems Versatile Irradiance Platform (VIP) — 79
4. Mercury and xenon short arc lamps

IV. REFERENCES — 82

CHAPTER IV

UV CURING EQUIPMENT - MONOCHROMATIC UV LAMPS

I. **MONOCHROMATIC UV RADIATION FOR CURING**   83

1. Emission from dielectric barrier discharges in rare gases and rare gas habide mixtures   84
   - (i) Dielectric barrier discharge
   - (ii) Rare gas excimers
   - (iii) Rare gas halide exciplexes (excimers)
   - (iv) Halogen excimers
   - (v) Mercury vapor excimers
   - (vi) General characteristics of incoherent excimer radiation
2. Commercial barrier discharge driven excimer lamps   94
3. Emission from microwave discharge powered excimer lamps   96
4. Commercial high pressure microwave excited excimer lamps   100

III. **EXCIMER LAMPS IN COMPARISON TO MEDIUM PRESSURE MERCURY LAMPS**   101

IV. **REFERENCES**   105

CHAPTER V

DOSIMETRY FOR EB AND UV CURING

I. INTRODUCTION

II LOW-ENERGY ELECTRON BEAM DOSIMETRY
    1. Interactions of Electrons with Matter and Dose   108
    2. Dosimetry with Thin-Film Dosimeters   112
       (i) Types, Properties and Evaluation of Dosimeters
       (ii) Calibration of Dosimeters
       (iii) Applications of Thin-Film Dosimeters

III. ULTRAVIOLET RADIATION DOSIMETRY   119
  1. Basic Terms, Quantities and Units
  2. Radiometers
  3. Chemical Actinometers
  4. Comparison of Radiometer with Chemical Actinometer
  5. UV-Sensitive Films and Labels

IV. REFERENCES   132

CHAPTER VI

ELECTRON BEAM (EB) CURING EQUIPMENT

I. GENERATION OF ACCELERATED ELECTRONS   135
1. Principle of electron acceleration
2. Characteristics of EB curing in comparison with thermal and UV curing
3. Definitions and units

II. TYPES OF INDUSTRIAL LOW-ENERGY ELECTRON PROCESSORS   142
1. Accessible electron energy and beam power range
2. Low-energy electron accelerators used in industry
3. Low-energy scanned beam electron accelerators
4. Linear cathode electron accelerators
5. Multi-filament Linear Cathode Electron Accelerators
6. Recent developments in electron beam curing equipment

III. REFERENCES   158

CHAPTER VII

UV CURING TECHNOLOGY - UV CURING UNITS AND APPLICATION TECHNIQUES

I. UV CURING UNITS - GENERAL DESIGN PRINCIPLES ... 159
1. Design Requirements
2. UV cure efficiency
3. Heat Management
4. System Reliability
5. Integration in a production line

II. UV CURING OF COATINGS AND INKS ON TWO-DIMENSIONAL SUBSTRATES ... 172
1. Curing of coatings or inks on flat, rigid substrates ... 172
   (i) Wood finishing
   (ii) UV Laminating for print finishing
   (iii) UV Screen printing
2. Curing of printing inks on flexible substrates ... 176
   (i) UV curing in offtet lithographic printing
       Web and sheet fed offset
   (ii) UV Curing in flexographic printing
   (iii) UV Curing in gravure printing
3. UV curing of coatings on flexible substrates ... 183
   (i) Coating methods
   (ii) UV Curing of silicon release coatings
   (iii) UV Curing of pressure sensitive adhesives
   (iv) UV Curing of lacquers, varnishes and paints
   (v) EB/UV Curing in cast coating

III. UV CURING OF INKS AND COATINGS ON CYLINDRICALLY SHAPED PARTS ... 189

IV. UV CURING OF THREE-DIMENSIONAL PARTS ... 190
1. Three-dimensional (3D) UV curing
2. Spot curing

V. UV MATTING OF COATINGS ... 192
VI. REFERENCES ... 194

CHAPTER VIII

RADIATION CURING TECHNOLOGY – UV CURING

I. DEGREE OF CURE AND CURE SPEED 195
1. Degree of cure
2. Evaluation of kinetic parameters in radical curing 196
    (i) Polymerisation rate
    (ii) Experimental determination of the degree of cure and the polymerisation rate
    (iii) The polymerisation rate as a function of irradiance
    (iv) The polymerisation rate as a function of photoinitiator concentration
    (v) Quantum yield of photoinduced polymerisation
    (vi) Effect of oxygen on photoinitiated polymerisation - the induction period
    (vii) Photochemical dark reaction (postcure)
    (viii) Experimental determination of propagation and termination rate constants
    (ix) Effect of temperature on the polymerisation rate and the degree of cure
    (x) Effect of temperature on the induction period
3. Cure speed 215
4. Chemical and physical factors affecting the cure speed 217
    (i) Effect of chemical factors
        (a) Function
        (b) Photoinitiator
        (c) Pigmentation
    (ii) Effect of physical factors
        (a) Irradiance in the curing plane
        (b) Oxygen inhibition
        (c) Nitrogen inerting
        (d) Temperature
        (e) Multiple exposure
5. Evaluation of kinetic parameters in cationic curing 226
    (i) Polymerisation rate
    (ii) The polymerisation rate as a function of irradiance
    (iii) The polymerisation rate as a function of photoinitiator concentration

	(iv)	Photochemical dark reaction (postcuring)	
	(v)	Effect of temperature on the polymerisation rate	
6.	Cure speed in cationic systems		234
7.	Chemical and physical factors affecting the cure speed of cationic systems		235

      (i)   Effect of chemical factors

           (a) *Addition of alcohols and vinylethers to epoxies*

           (b) *Photoinitiators*

           (c) *Water and nucleophiles*

      (ii)   Effect of physical factors

           (a) *Temperature*

           (b) *Oxygen effect and nitrogen inerting*

8.	Evaluation of kinetic parameters in electron beam curing		237
	(i)	Mechanism of electron beam curing	
	(ii)	Polymerisation rate at radical initiation	
	(iii)	Effect of oxygen in electron beam curing	
	(iv)	Pigmentation in electron beam curing	
II.	**DEGREE OF CURE: PHYSICAL AND CHEMICAL CHARACTERISATION**		**243**
1.	Methods to measure the degree of cure		243

      (i)   Field cure tests

      (ii)   Laboratory cure test methods

           (a) *Infrared and Raman spectroscopy*

           (b) *Real-time infrared spectroscopy*

           (c) *Confocal Raman spectroscopy*

           (d) *Photo- Calorimetry (DSC)*

           (e) *Fluorescence probe technique*

           (f) *Photoacoustic FT IR spectroscopy*

           (g) *Gel Content and Extractables*

           (h) *Dynamic Mechanical Analysis (DMA) and Rheometry*

           (i) *Dilatometry*

III. REFERENCES   263

CHAPTER IX

UV&EB EQUIPMENT HEALTH AND SAFETY

1. **UV Equipment Health and Safety**   265
   - (a) *Photobiological effects of UV radiation*
   - (b) *Shielding for UV equipment*
   - (c) *Shielding for heat*
   - (d) *Protection against ozone*
   - (e) *High voltage protection*
2. **EB Equipment Health and Safety**   271
   - (a) *X-ray formation by fast electron*
   - (b) *Biological action of x-rays*
   - (c) *Radiation shielding of low-energy electron accelerators*

REFERENCES   278

CHAPTER I

RADIATION CURING: DEFINITION AND BASIC CHARACTERISTICS

I. INTRODUCTION

1. Definition of the term radiation curing

The transformation of a reactive liquid into a solid by radiation, leading to polymerisation and in most cases also to cross-linking is termed curing. Infrared, microwave and radio frequency radiation initiates thermal curing, whereas ultraviolet (UV) and electron beam (EB) irradiation via electronic excitation and ionisation lead to non-thermal curing. This volume of the series "Chemistry & Technology of UV & EB Formulation for Coatings, Inks & Paints" concentrates on UV and EB radiation curing, The term radiation curing is used in this sense.
UV (EB) radiation curing is defined as:-

> *the fast transformation of 100% reactive, specially formulated liquids into solids by UV photons or electrons.*

This simple definition refers to the well-known fact that organic molecules become electronically excited or ionised after energy absorption. Energies between 2 and 8 eV are typically needed to transform organic molecules from the ground to an excited state. The excited molecules are able to undergo chemical reactions which may lead to chemically reactive products, which initiate the fast transformation of the liquid into a solid.

Ionisation of organic molecules occurs at higher energies. As a result of the ionisation process positive ions and secondary electrons are generated. In the case of liquid acrylates as 100% reactive liquids, positive ions are transformed into radicals. Secondary electrons lose their excess energy, become thermalised and add to the acrylate. The radical anions formed are a further source of radicals capable of inducing fast transformation.

In industrial radiation curing applications either electrons with energies between 100 and 300 keV or UV photons with energies from 2.2 to 7.0 eV are applied.
Fast electrons transfer their energy to the molecules in the "reactive liquid" during a series of electrostatic interactions with the outer sphere electrons of the neighbouring molecules. This leads to excitation and ionisation and finally to the formation of chemically reactive species.

In contrast to electrons, photons are absorbed by the chromophoric site of a molecule in a single event. In UV curing applications special photoinitiators are used which absorb the photons and generate radicals or protons. The fast transformation from the liquid into solid takes place as radical or cationic polymerisation which in the most cases is accompanied by cross-linking. Fast transformation means that the polymerisation process mainly proceeds within a time range of 10^{-2} - 1 s. However, in a rigid polymer matrix radicals or cationic species can live much longer than a few seconds. A post or dark cure process proceeds after irradiation. As an example, a schematic representation of radical formation, polymerisation and cross-linking is shown in Figure 1.

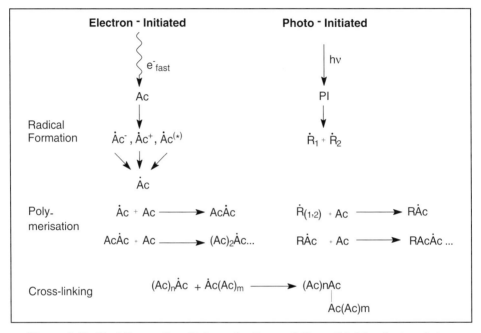

Figure 1 Radical Formation, Polymerisation and Cross-Linking in Acrylates
(R = Radical, Ac = Acrylate, PI = Photoinitiator)

As a result of the curing process, a solid polymer network is formed from a 100% reactive liquid. The latter term means that the liquid radiation curable system does not contain any main components, which do not take part in the polymer formation. Especially, as in these formulations, volatile organic solvents are avoided. Radiation curable formulations are solvent-free and consist of 100% reactive monomers and oligomers[1,2,3,4,5,6].

II. CHEMICAL SYSTEMS USED FOR RADIATION CURING

1. Acrylate and methacrylate systems

The most commonly used UV or electron beam (EB) curable formulations contain the acrylate unsaturation $H_2C=CR-COOR`$ (R=H: acrylate, R=CH_3: methacrylate). Methacrylates are less reactive than acrylates and are used in special applications.

The polymerisation of acrylates is initiated by radicals. Chain propagation and termination as well as cross-linking are radical processes. The polymerisation of acrylates is a typical radical polymerisation of vinyl monomers [7]. Acrylate double bonds are consumed during the polymerisation process. Therefore, the degree of double bond conversion is a measure of the degree of cure. Using experimental methods such as stationary or time-resolved infrared spectroscopy, the double bond conversion can be either determined after a certain time or followed as a function of time [8].

A typical time profile of the double bond conversion of an UV irradiated acrylate containing a photoinitiator is shown in Figure 2.

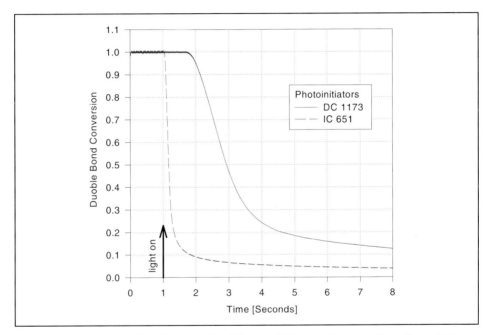

Figure 2 The UV Curing Process: Double Bond Conversion as a Function of Time

The most important performance characteristics which have to be met by solvent free radiation curable acrylate formulations are high reactivity and adjustable viscosity. Rapid cure speed and low viscosity are obtained when acrylate monomers are used. However, in most cases formulations containing only monomers show coating properties which are totally unacceptable. Brittleness, bad adhesion to the substrate and a high content of extractable substances are typical characteristics of such coatings. On the other hand, acrylate oligomers are available, which usually have higher viscosity and lower reactivity than acrylate monomers but are to be meet a broad range of coating property requirements. Therefore, radiation curable formulations usually consist of monomers as reactive thinners and oligomers as binders.

Multifunctional acrylate monomers based on various polyols such as TMPTA, TPGDA, HDDA, PETA etc. (see Table I) have been used since the early 70's. These monomers show good diluency in combination with high cure speed and low volatility. They are still widely used. However, there is an ongoing trend towards the development of materials less likely to irritate the skin. Based on alkoxylated polyols, numerous acrylate monomers are now available in this category.

Table I Bifunctional and Polyfunctional Acrylates

Tripropylene glycol diacrylate	TPGDA
1,6-Hexanediol diacrylate	HDDA
Dipropylene glycol diacrylate	DPGDA
Trimethylolpropane triacrylate	TMPTA
Trimethylolpropane ethoxytriacrylate	TMP(EO)TA
Trimethylolpropane propoxytriacrylate	TMP(PO)TA
Pentaerythritol triacrylate	PETA
Glyceryl propoxytriacrylate	GPTA

The main types of acrylic oligomers are epoxy acrylates, polyester acrylates, polyether acrylates, urethane acrylates and silicone acrylates. The chemical structures of some oligomers mentioned are given in Table II.

Table II Acrylate Oligomers

Polyester acrylates

$$CH_2=CH-\overset{O}{\underset{\|}{C}}-O-(CH_2)_6-[O-\overset{O}{\underset{\|}{C}}-(CH_2)_4-\overset{O}{\underset{\|}{C}}-O-(CH_2)_6]_n-O-\overset{O}{\underset{\|}{C}}-CH=CH_2$$

Epoxy acrylates

$$CH_2=CH-\overset{O}{\underset{\|}{C}}-O-CH_2-\overset{OH}{\underset{|}{CH}}-R-\overset{OH}{\underset{|}{CH}}-CH_2-O-\overset{O}{\underset{\|}{C}}-CH=CH_2$$

Examples:

$$R = -CH_2-O-\bigcirc-\underset{\underset{CH_3}{|}}{\overset{\overset{CH_3}{|}}{C}}-\bigcirc-O-CH_2-$$

Table II Acrylate Oligomers (continued)

Polyurethane acrylates

$$CH_2=CH-\overset{O}{\overset{\|}{C}}-O-(CH_2)n-O-\overset{O}{\overset{\|}{C}}-\overset{H}{\overset{|}{N}}-R-\overset{H}{\overset{|}{N}}-\overset{O}{\overset{\|}{C}}-O-(CH_2)n-O-\overset{O}{\overset{\|}{C}}-CH=CH_2$$

Examples:

R = (4-methylphenyl with NCO), OCN—(2-methylphenyl)—NCO ; OCN—$(CH_2)_6$—NCO

Silicone acrylates

$$CH_2=CH-\overset{O}{\overset{\|}{C}}-O-R-\underset{CH_3}{\overset{CH_3}{\underset{|}{\overset{|}{Si}}}}-[-O-\underset{CH_3}{\overset{CH_3}{\underset{|}{\overset{|}{Si}}}}-]n-O-\underset{CH_3}{\overset{CH_3}{\underset{|}{\overset{|}{Si}}}}-R'-O-\overset{O}{\overset{\|}{C}}-CH=CH_2$$

Examples:

R, R' = —CH_2—C=CH_2 , —CH_2—CH=CH—

Polyether acrylates

$$CH_2=CH-\overset{O}{\overset{\|}{C}}-O-(CH_2)n-O-CH-O-(CH_2)n-O-\overset{O}{\overset{\|}{C}}-CH=CH_2$$
$$|$$
$$(CH_2)n-O-\overset{O}{\overset{\|}{C}}-CH=CH_2$$

Aromatic and aliphatic epoxy acrylates are the most widely used group of radiation curable oligomers despite the fact that they have to be thinned with monomers.

Epoxy acrylates can be synthesised using a catalytic reaction of the epoxy group of e.g. bisphenol A diglycidyl ether with either acrylic or methacrylic acid. Epoxy acrylates are highly reactive and produce, typically hard and chemically resistant coatings. This makes them suitable for wood finishing applications, varnishes for paper and cardboard as well as for hard coatings.

Urethane acrylates are formed by reacting isocyanates with hydroxy functional acrylate monomers. The incorporation of polyols, polyester polyols or polyether polyols into urethane acrylates leads to a variety of modified structures. Flexible films are obtained when long chain glycols are used to modify the urethane moeity. Hard films can be produced when highly branched multifunctional polyols are employed. As a class of compounds produced by a versatile chemistry, urethane acrylates offer a wide choice of excellent application properties. Flexibility, abrasion resistance, toughness, good adhesion to difficult substrates, chemical resistance and excellent weathering resistance are film properties that can be obtained by a suitable selection of aromatic or aliphatic urethane acrylates.

Polyester acrylates are either produced by reacting the OH group of polyesters with acrylic acid or hydroxy acrylate with acid groups of the polyester structure. Polyester acrylates are often low-viscosity resins requiring little or no monomer. Amino modified polyester acrylates especially show high reactivity and low viscosity as well as low skin irritation, in some cases even meeting the requirements of Xi free materials. They are also useful as amine synergists or coinitiators in UV curable systems providing abstractable hydrogen for hydrogen abstracting photoinitiators.

Polyester acrylates cover a wide range of viscosity and reactivity. They are often applied in wood coatings, varnishes, lithographic and screen inks. Additionally, purified polyester acrylates are available which have the benefit of lower extractables and odour.

Polyether acrylates are produced by esterifying polyetherols with acrylic acid. They can reach even lower viscosities than polyester acrylates and do not require reactive thinners. Amino modification of polyether acrylates leads to similar benefits in reactivity and low skin irritance as those offered by polyester acrylates.

Silicone acrylates are acrylated organopolysiloxanes. As the most important class of silicone acrylates the backbone consists of polydimethylsiloxane units. The silicone backbone is responsible for flexibility and resistance to heat, moisture, radiation degradation and shear forces. Good release properties are due to the low intermolecular interactions induced by the methyl group. Reactivity, degree of cure and release force are adjustable over a wide range by variation of the acrylate functionality and the steric arrangement of the reactive groups within the polymer chain.

Some polyester, urethane and epoxy acrylates are either water soluble, or dilutible, or can be emulsified using water and emulsion stabilisers. Water-based acrylate systems have a reduced viscosity, show lower odour and skin irritation, good adhesion and matting properties. They avoid some drawbacks of conventional polymer dispersions such as moderate chemical resistance, poor mechanical strength and low gloss of the coatings produced. However, the water has to be evaporated prior to, or during irradiation. Higher energy consumption, susceptibility to humidity and problems with the release of organic harmful substances is another potential disadvantage of water-based systems. These drawbacks are partially removed by using air drying urethane acrylates. Their films exhibit good blocking resistance even before cross-linking.

Solid urethane or polyester acrylates can be main components of radiation curable powders. Together with suitable unsaturated polyesters, powders are formed which give low film flow temperatures and allow separating film formation from curing. Using this technology powder coatings for wood and plastics become feasible [9].

2. Cationic systems

The cationic polymerisation of substances containing oxirane functionality such as aliphatic and cycloaliphatic epoxides as well as glycidyl ethers can be initiated by photogenerated Brønsted or Lewis acids [10]. However, cationic UV curing is not restricted to epoxides. Vinyl ethers can also polymerise by a cationic mechanism and form an interpenetrating polymer network if blended with epoxides. The addition of vinyl ethers increases the cure speed of epoxides and decreases viscosity. Multifunctional polyols are frequently used in cationic UV curing formulations as chain transfer agents, improving the cure speed, modifying the cross-link density and affecting the flexibility of the coating.

Some important constituents of cationic UV curing systems are summarised in Table III.

Table III   Expoxides, Vinyl Ethers, Glycidyl Ethers and Epoxysilicones Used in Cationic UV Curing Systems

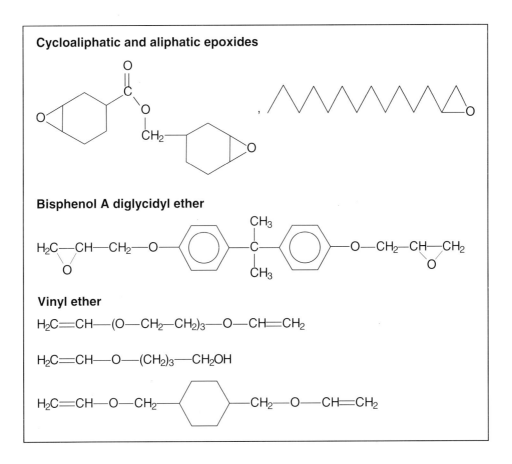

**Table III Epoxides, Vinyl Ethers, Glycidyl Ethers and Epoxysilicones
Used in Cationic UV Curing Systems** (continued)

Epoxy - functional silicone polyether block copolymer

[Chemical structure of epoxy-functional silicone polyether block copolymer]

Epoxy - functional poly dimethylsiloxane oligomers

[Chemical structure of epoxy-functional poly dimethylsiloxane oligomers]

Cationic UV curing formulations can be used for coatings showing a wide range of properties, including good adhesion, impact resistance, chemical resistance, hardness and abrasion resistance. There are certain characteristics of cationic curing which are of general importance: The cure speed of cationic UV curing systems is generally lower than that of acrylates. Cationic cure is not affected by oxygen but negatively affected by ambient humidity, amines or other bases, including certain pigments. Heat will increase the cationic cure speed. In some cases additional heating is applied to reach a high degree of conversion. The photogenerated acids (protons) continue to be active after UV exposure. A pronounced dark (post) cure effect is observed which can even last days after UV irradiation.

Epoxysilicones are applied as release coatings. Special epoxysilicone polyether block copolymers show good photoinitiator miscibility, high cure speed and compatibility with epoxy and vinyl ether monomers. These block copolymers form flexible films with excellent release properties.

3. Maleate/vinyl ether systems

Maleate/vinyl ether (MA/VE) UV curing systems have been commercially used since 1990 [11]. They contain only oligomeric components, are not skin irritating nor sensitizing and exhibit a low vapour pressure, i.e. in coating applications the concentration of volatile organic compounds is kept very low. However, their cure speed is generally lower than for arylate formulations. The MA/VE systems consist of maleate and vinyl ether double bonds with a 1:1 stoichiometric ratio. When radicals are generated, e.g. from conventional radical photoinitiators, copolymerisation occurs via an electron donor acceptor complex. For practical applications, oligomeric MA/VE systems have been developed which consist of maleate end caps and a polyester, epoxy or urethane backbone section which is diluted by means of a vinyl ether. This is schematically shown in Figure 3.

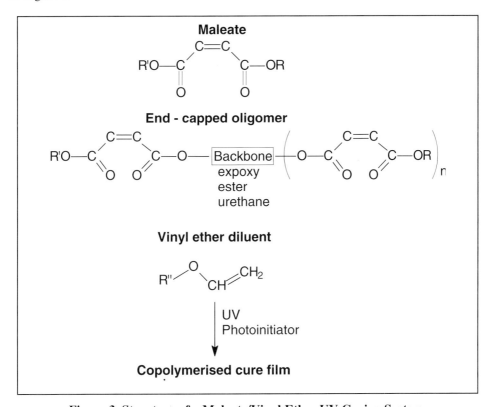

Figure 3 Structure of a Maleate/Vinyl Ether UV Curing System

At present MA/VE systems are mainly used for wood coatings in roll and spray applications. Mixing with acrylates is possible and further extends formulation latitude. MA/VE systems can also be used as binders for UV curable powder coatings. The binders contain two polymers: an unsaturated polyester containing maleic and fumaric acid functionalities, and a polyurethane with vinyl ether unsaturation.

4. Donor acceptor complexes

If unsaturated molecules containing an excess electron charge at the double bond, such as vinyl ethers or styryloxy derivatives, are mixed with maleic anhydride, maleimides, maleates or fumarates, which have an electron deficient unsaturation, donor acceptor complexes (DAC) can either be formed during mixing (ground state DAC) or after UV irradiation (excited state DAC). Some selected acceptors and donors are shown in Figure 4 [12].

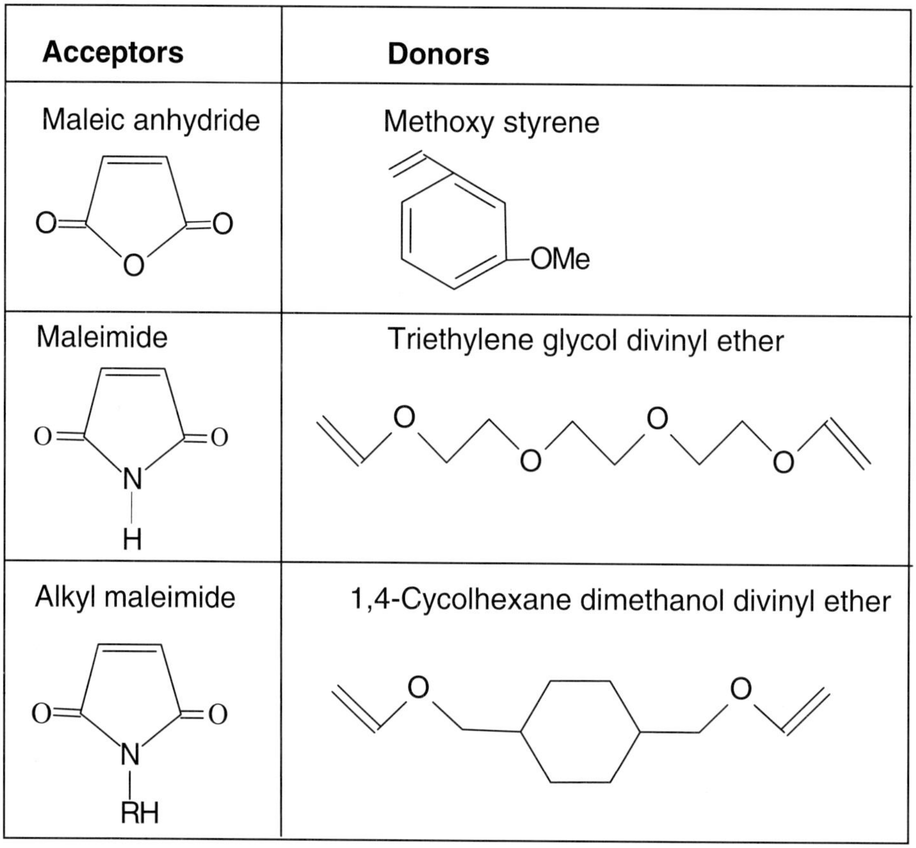

Acceptors	Donors
Maleic anhydride	Methoxy styrene
Maleimide	Triethylene glycol divinyl ether
Alkyl maleimide	1,4-Cycolhexane dimethanol divinyl ether

Figure 4 Donor Acceptor Pairs

The excited DAC probably undergoes H-abstraction or electron transfer and acts as a radical source for subsequent copolymerisation. It is of particular interest that some electron acceptors form strong chromophores for UV photon absorption thus playing the role of a photoinitiator radical source.

The DAC systems mentioned offer a potential as non-acrylate, non-photoinitiator UV curing systems.

5. Unsaturated Polyesters

Unsaturated polyesters belong to the earliest commercially available radiation curable systems. They are condensation products of organic diacids and glycols. The structure of a typical unsaturated polyester is schematically shown in Figure 5. Phthallic and/or maleic anhydride and 1,2 propylene glycol are used to prepare unsaturated polyesters in a condensation reaction.

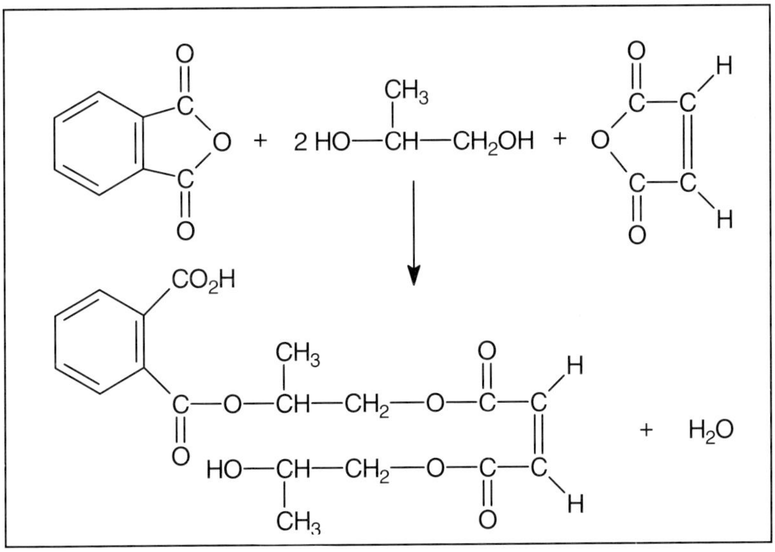

Figure 5 Formation Scheme of an Unsaturated Polyester

To reduce the viscosity of the unsaturated polyester 20 to 40 mass% of styrene is usually added. Styrene/polyester systems are free radical cured. In comparison to acrylate systems they cure at a slower rate, and the network cross-link density is lower.

Styrene/polyester systems still find widespread usage as wood coatings. They are used because of their low costs, but acrylate systems are believed progressively to replace them. The essential reasons for this substitution is to avoid volatile styrene, to develop low viscosity sprayable systems, to increase reactivity and line speed and to improve the mechanical characteristics of the coatings.

6. Thiol-ene systems

Thiol-ene systems are composed of a mixture of polythiols and olefinic compounds. The addition of thiols to the olefinic double bond can occur by either radicals or ions. Despite the fact that their cure is not inhibited by oxygen these systems do not find wide application. The odour of volatiles emitted from polythiol compounds is a reason for their restricted application.

III. INITIATION OF CURING: ACTION OF PHOTONS AND ELECTRONS

1. Physical characteristics of EB and UV sources

In industrial radiation curing applications either electrons with energies between 100 and 300 keV or photons with energies ranging from 2.2 to 7.0 eV are applied. The mercury discharge, as used in mercury arc lamps, produces a polychromatic spectrum with intensive emission lines exhibiting energies from 2.8 to 6.0 eV. Monochromatic UV radiation is emitted from excimer lamps. In this case a microwave discharge [13] or a radio frequency driven silent discharge [14] generates excimer excited states of noble gas or noble gas halide molecules, which decay by the emission of monochromatic UV radiation. In the ground state, the excimer molecules decay into atoms. Therefore, no self-absorption of the UV radiation can occur. All photons are coupled out of the discharge. At present, emission from XeCl (308 nm = 4.02 eV) and KrCl (222 nm = 5.58 eV) excimers reaches power levels required for fast cure speeds. Photons with an energy of 7 eV (172 nm) as emitted from Xe_2 excimers are applied for special surface cure applications, whereas CuBr excimer radiation (2.2 eV = 535 nm) allows visible light curing.

The efficiency of different radiation sources, defined as the ratio of the output electron beam power to the input sum electrical power in the case of electrons, or the useful total radiant power to the electrical power of the lamp, show great variation. Low-energy electron accelerators reach maximum efficiencies of 60%. Advanced mercury medium pressure lamps transform up to 30% of the electric input lamp power into UV radiation, and excimer lamps up to 12 % into (UVorVIS) photon output power. Typical power densities as measured in the curing position are 10 - 40 Wcm^{-2} for electrons, 1-3 Wcm^{-2} for mercury lamps and 0.1-5 Wcm^{-2} for XeCl lamps. Some basic physical characteristics of electron and UV sources, as used for curing applications, are summarised in Table IV.

Table IV Basic Physical Characteristics of EB and UV Sources

Feature	Electrons	Photons
Energy	100 – 300 keV	polychromatic: 6.2 – 2.7 eV
		monochromatic: 2.2, 4.0, 5.6, and 7.0 eV (further excimer transitions available)
Fluence or Irradiance	10-40 Wcm^{-2}	polychromatic: 1-3 Wcm^{-2} monochromatic: 0.1-0.3 Wcm^{-2} (RF powered) 1-5 Wcm^{-2} (microwave powered)
Energy absorption	in bulk	in photoinitiator

2. Absorption of electrons and photons in matter

When electrons with typical energies in the keV and MeV range are absorbed in matter, secondary electrons are produced as a result of the energy degradation process. These enter into electrostatic (Coulomb) interaction with atoms or molecules of the absorber, which finally results in the formation of ions, radicals and excited states of the absorber molecules or atoms. Thus, electrons are capable of inducing chemical reactions in materials. The material is modified by the action of fast electrons. Electron beam curing of liquids containing unsaturations can be regarded as a special case of this modification.

Chemically reactive species which can initiate curing are generated within the penetration range of the fast electrons (see Chapter VI Figure 5). The resulting energy absorption profile is represented by the corresponding depth dose distribution. Using a logarithm developed by Tabata and Ito [15], the depth dose profiles shown in Figure 6 were calculated for electrons with energies from 100 to 200 keV, absorbed in a material with a density of 1 gcm^{-3}, an atomic number of 6 and an atomic weight of 12, conditions which are approximately fulfilled in the electron beam curing of organic liquids. In Figure 6 the energy loss dE/dx of the fast electrons (expressed in units MeV cm^{-1}) is given as a function of the mass per unit area (in gcm^{-2}). The energy loss is directly proportional to the electron absorbed dose. This can be shown using the relationship (ρ = density of the absorber) $1/\rho \: dE/dx$ = dose x unit area, or in dimensions Jg^{-1}cm^{2}. Per definition, the dose of 1 Jg^{-1} is 1 kGy.

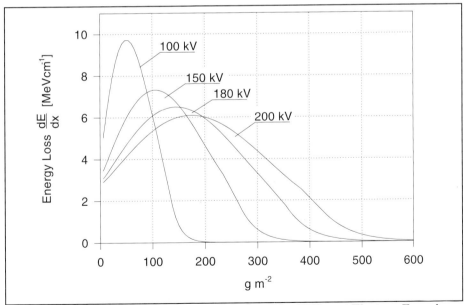

Figure 6 Depth Dose Distributions Calculated for Different Electron Energies

It is sometimes assumed that, geometrically, the depth dose profile corresponds to the degree of cure profile. However, this conclusion has to be regarded with caution. Radicals and ions involved in the curing process can live much longer than the exposure time. It was experimentally observed that radical sites and protons are able even to migrate over macroscopic distances. Thus, the cure depth profile and depth dose profile can differ.

Contrary to electrons, photons are absorbed by the chromophoric site of a molecule in a single event. In UV curing application, the chromophore is part of the photoinitiator molecule. In general, the photon absorption follows Lambert-Beer`s law. This is shown in Figure 7. The number of photons I present at depth l from the surface is given as a function of the optical absorbency A = εx[PI]xl normalised to the initial number of photons I_0. The optical absorbance A is expressed as the product of the photoinitiator concentration [PI], the molar extinction coefficient (at the wavelength absorbed) ε of the photoinitiator and the photon penetration path length l.

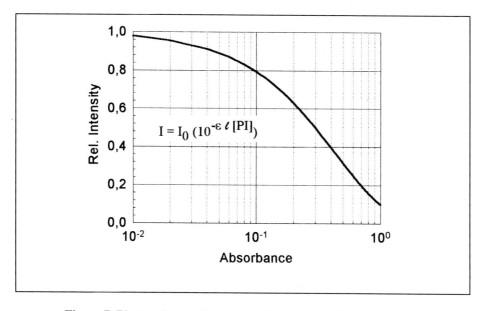

Figure 7 Photon Absorption According Lambert Beer`s Law

Lambert-Beer`s law is used here to illustrate some general points. It cannot be applied for reflected or scattered light.

The depth profile of rapidly formed photoinitiator generated radicals or ions corresponds to the inverse photon penetration profile. As in the case of electron initiation, the final cure profile can deviate from the initial radical distribution.

3. Initiation of UV curing by free radicals

The UV curing of acrylate, methacrylate and maleate/vinyl ether systems is initiated by free radicals. In all practical cases the initiating radicals are generated from electronically excited photoinitiator molecules [16,17]. As shown schematically in Figure 8, a photoinitiator molecule AB is excited into the singlet state by photon absorption.

Radical formation occurs via a triplet state. Fast transformation of the singlet into a triplet by intersystem crossing (ISC) is a necessary condition for obtaining a high radical yield. Side reactions such as singlet decay by fluorescence or triplet quenching by oxygen have to be avoided. Radical formation occurs via two possible reaction sequences. They are designated as Norrish Type I and Type II reactions. In the first case the photoinitiator triplet state decays into a radical pair by homolytic decomposition. A geminate radical pair is formed surrounded by a solvent cage. Diffusional motion prevents the immediate recombination of the radical in the pair and is the driving force for the escape of radicals from the cage. Under normal conditions only these "free radicals" can initiate the polymerisation.

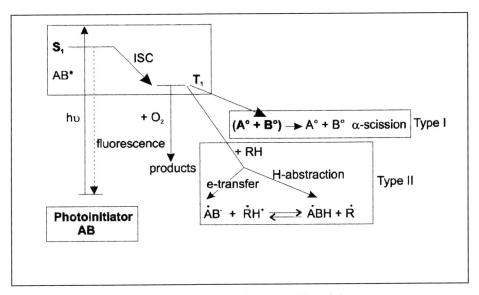

Figure 8 Radical Formation from Photoinitators

Triplet states of benzophenone and of some benzophenone derivatives preferably react with suitable hydrogen donating compounds RH by hydrogen abstraction. The resulting radical pair can either be generated by a homolytic cleavage of the R-H bond or via an intermediate charge transfer complex followed by proton transfer.

Some examples of Type I and Type II photoinitiators are shown in Figures 9a & b.

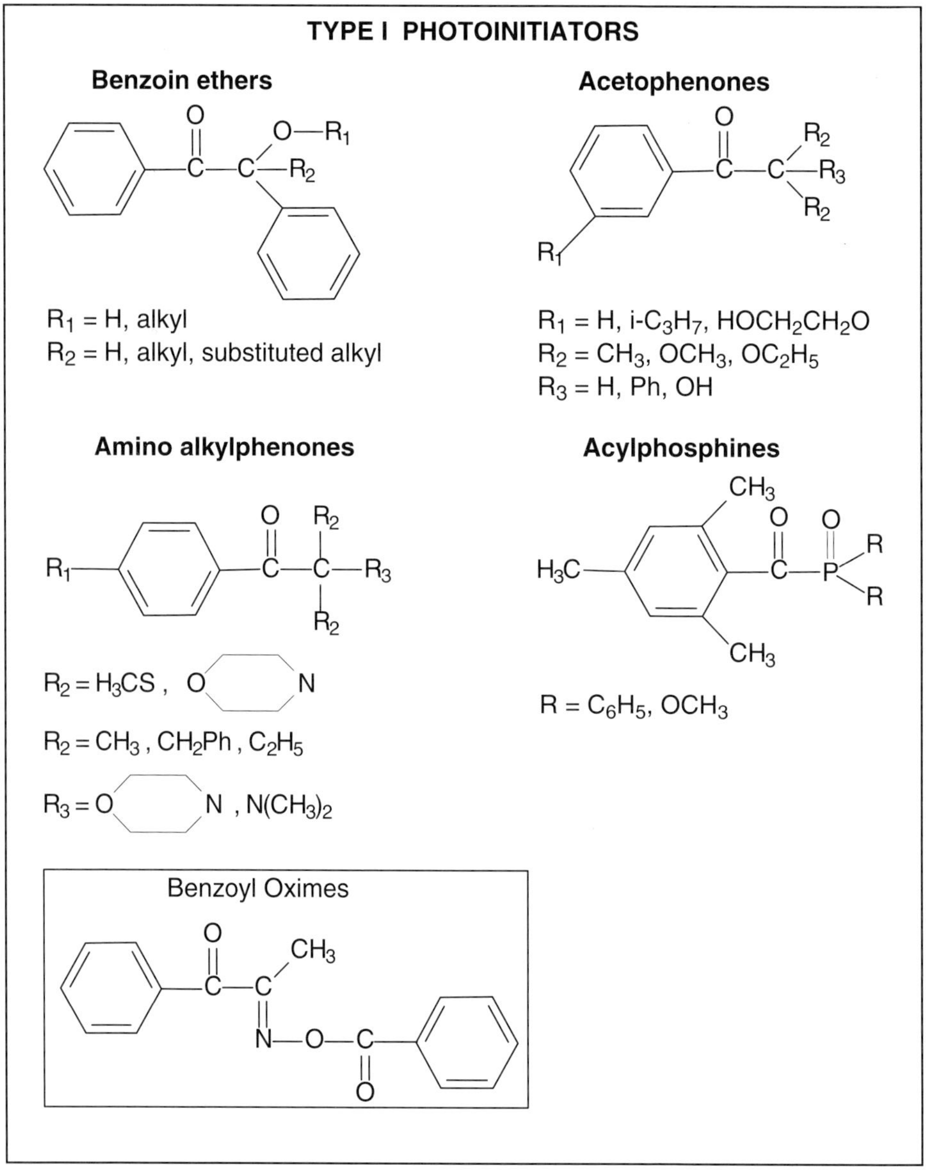

TYPE I PHOTOINITIATORS

Benzoin ethers

R_1 = H, alkyl
R_2 = H, alkyl, substituted alkyl

Acetophenones

R_1 = H, i-C_3H_7, $HOCH_2CH_2O$
R_2 = CH_3, OCH_3, OC_2H_5
R_3 = H, Ph, OH

Amino alkylphenones

R_2 = H_3CS, O⌐⌐N

R_2 = CH_3, CH_2Ph, C_2H_5

R_3 = O⌐⌐N, $N(CH_3)_2$

Acylphosphines

R = C_6H_5, OCH_3

Benzoyl Oximes

Figure 9a Examples of Type I Photoinitiators

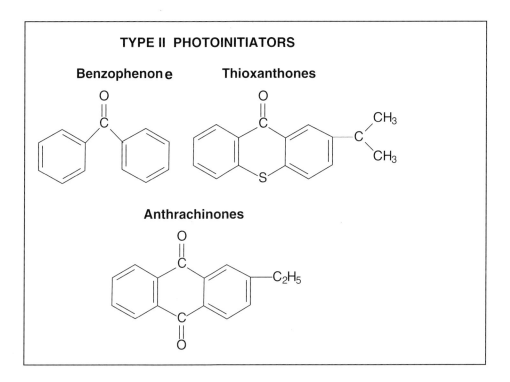

Figure 9b Examples of Type II Photoinitiators

Photoinitiated free radical polymerisation proceeds via three main reaction steps: initiation, propagation and termination. As shown in Figure 10, the rate of initiation v_i is proportional to the product of the radical yield per absorbed photon Φ_i and the number of photons absorbed per second I_a. I_a is a fraction of I_0, the number of photons per second entering the process zone.

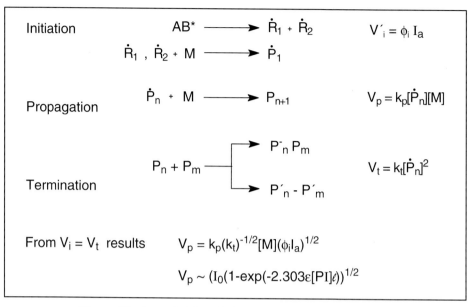

Figure 10 Steps in Photoinitiated Radical Polymerisation

Chain propagation mainly consumes the monomer. The propagation rate depends on the monomer concentration [M] the concentration of polymer radicals [$P_n°$] and the propagation rate constant k_p. Termination occurs by combination or disproportionation of different polymer radicals. The termination rate v_t is proportional to the polymer radical concentration [$P_n°$] squared. Further chain termination processes can be chain transfer and reaction of polymer radicals with inhibitors.

It should be pointed out that it is only the initiation step, radical formation from the photoinitiator, which is different from thermal radical polymerisation.
In order to estimate the polymerisation (propagation) rate v_p the following assumptions have to be made:

- monochromatic light is used, which is exclusively absorbed by the photoinitiator,
- absorption is small and homogeneous in the volume irradiated,
- when polymerisation proceeds, a stationary radical concentration is obtained,
- all polymer radicals show the same reactivity towards propagation and termination.

Using this simplified picture the expression given in Figure 10 is obtained for the polymerisation rate v_p.

Initiation of curing via photoexcited donor acceptor complexes (see Figure 4) can be regarded as a special case of radical polymerisation. The ground state complex itself or the acceptor acts as chromophore. The excited donor acceptor complex undergoes hydrogen abstraction or electron transfer, thus generating the initiating radical pair.

With the development of UV irradiation sources such as 172 and 222 nm excimer lamps, the free radical polymerisation of photoinitiator-free neat acrylates becomes feasible. Photons emitted from these lamps are absorbed by many acrylates, thereby directly generating radicals.

4. Initiation of UV curing by cations

Cationic polymerisation is either initiated by strong Lewis acids such as BF_3 or PF_5 or Brönsted acids such as $H^+BF_4^-$, $H^+PF_6^-$ or $H^+SbF_6^-$. After UV illumination aryldiazonium salts generate Lewis acids whereas diaryliodonium, triarylsulphonium and triarylselenium salts produce strong Brönsted acids. The latter are preferably used as initiating species in cationic polymerisation [10]. Structures of some commercially used arylsulphonium salts are given in Figure 11.

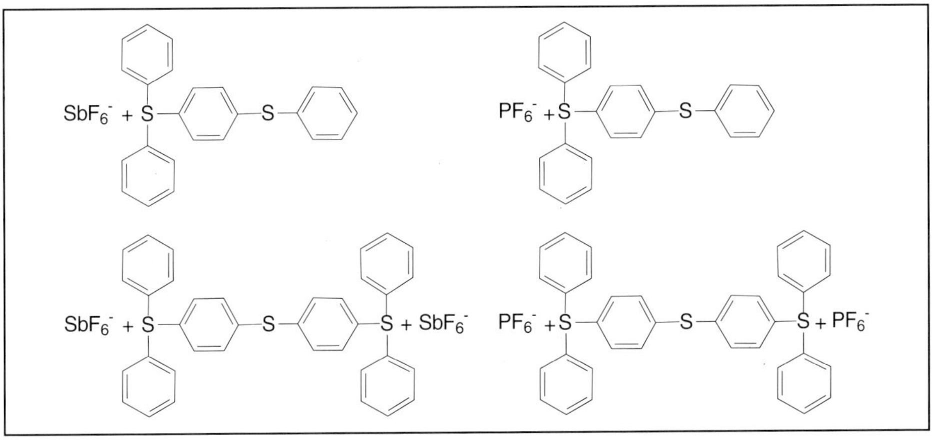

Figure 11 Structures of Arylsulphonium Salts

As schematically shown in Figure 12 direct photoinitiation of triarylsulphonium salts via homolytic or heterolytic cleavage of the sulphur hydrogen bond leads to the formation of solvent caged pairs containing either the phenyl radical (Ar_2S^+ + $Ar^•$) or the phenyl cation (Ar_2S + Ar^+). Escape from solvent cage and hydrogen abstraction result in the formation of the Brönsted acid HX. According to that pathway, diphenylsulphide (Ar_2S) and acid should be produced in equimolecular amounts. This is in disagreement with experimental results which showed a 3:1 excess in acid formation. Cage products are assumed to be the source of this "excess acid" Cage recombination can either yield the triarylsulfonium salt or via molecular rearrangement three phenylthiobiphenyl isomers (Ar_2SAr) and the corresponding acid, thereby explaining the excess amount of acid.

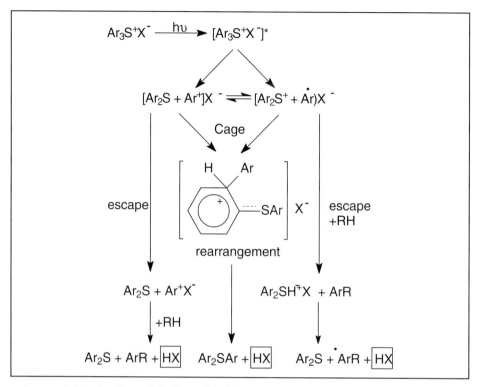

Figure 12 Mechanism of Brönsted Acid Formation from Triarylsulfonium Salts

The polymerisation of an epoxy monomer M is intitated by intermediate protonation of the oxirane ring followed by carbocation (HM^+) formation. Polymerisation proceeds via oxirane ring opening by reactive cations ($H-M_n-M^+$) and can be terminated by anionic or nucleophilic species. Therefore, anionic counterions (X^-) with low nucleophilicity should be selected for the initiating acid. If hydroxy functional compounds (ROH) are added, chain transfer can take place via proton formation. Initiation, propagation, chain transfer and termination of cationic polymerisation are schematically summarized in Figure 13.

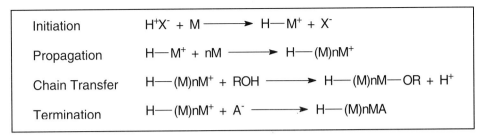

Figure 13 Reaction Steps in Photoinduced Cationic Polymerisation

Additionally, the photochemical decomposition of cationic photoinitiators can be sensitised by energy or electron transfer from the excited state of the sensitiser. Sensitisers such as isopropylthioxantone (ITX), anthracene, perylene and even dyes absorbing in the visible region can be used to shift excitation to higher wavelengths. This provides a method to utilise visible light, e.g. from tungsten lamps, for photocuring.

UV initiated cationic polymerisation differs from photoinduced free radical polymerisation in the following ways:

- the initiating species is a stable compound only consumed by anions or nucleophiles,
- after UV exposure, cationic polymerisation continues for a long time,
- since no radicals are involved, cationic polymerisation is not sensitve to oxygen,
- cationic systems are sensitive to humidity
- in general, the cure speed of cationic systems is lower than that of radical systems
- films made from cationic formulations show low shrinkage and good adhesion to numerous substrates.

5. Initiation of UV hybrid curing

Curing of formulations containing components which undergo polymerisation by two different mechanisms is called hybrid curing [18]. Photodecomposition of arylsulphonium salts generates both radicals and Brönsted acids. For example, the radicals are able to start the polymerisation of acrylates, and the acid that of vinyl ethers. In acrylate systems containing a cationic photoinitiator acrylate polymerisation is considerably accelerated by the addition of vinyl ethers. On the other hand, the vinyl ether component is cured cationically. Interpenetrating polymer networks are produced, containing both the acrylate and vinyl ether component. Vinyl ethers as good hydrogen donors enhance photoinduced radical production from aryl sulphonium initiators, thus increasing the initition rate of acrylate polymerisation.

In comparison with cationic curing systems, hybrid systems offer a higher cure speed, an improved solvent resistance of the cured films and a greater formulation latitude. Compared with radical curing systems, better adhesion to critical substrates and lower oxygen sensitivity is observed.

6. Initiation of electron beam curing: Free radical polymerisation

Electron beam curing is mainly used for acrylate formulations. As in the case of photoinitiation, the electron-initiated polymerisation of acrylates proceeds via free radicals. After the electron irradiation of liquid acrylates radicals are generated, which initiate polymerisation and form reactive sites for cross-linking. Typical irradiation doses of 10 to 50 kGy are needed to transform liquid di- and multifunctional acrylates into dense polymer networks. This curing process is used to produce functional coating with a variety of interesting properties and has found wide application.

Using pulse radiolysis and laser photolysis with optical detection as well as electron spin resonance (ESR) the mechanism of electron-induced radical formation in acrylate solutions was investigated. Figure 14 shows a reaction scheme, which summarises the main reaction pathways leading to polymerisation and cross-linking of acrylates such as TPGDA [19].

Irradiation of a liquid acrylate with fast electrons generates radical cations and solvated electrons as primary species. Acrylate radical cations deprotonate (Reaction 1) forming vinyl-type radicals. Vinyl-type radicals are able to react with acrylates and to initiate polymerisation (Reaction 2). If oxygen is present, vinyl-type radicals form vinyl peroxyl radicals exhibiting characteristic visible absorption bands (Reaction 5). Acrylate radical cations can also dimerise in a fast reaction generating distonic or resonance stabilized dimer cations (Reaction 3). It is assumed that distonic dimer cations also contribute to polymerisation. ESR experiments suggest that a radical transformation takes place. A carbon centered radical is generated, which either forms cross-links by recombination or chain growing radicals by addition to an acrylate.

Secondary electrons as produced by the fast primary electron become thermalised and solvated within picoseconds. The addition of solvated electrons to acrylates forms acrylate anions (Reaction 6). Radical anions of difunctional acrylates are capable of forming resonance stabilised dimer anions (reaction 11).

Protonation of the anions at the carbonyl oxygen leads to ketyl-type radicals (Reaction 7). Protonation of the anions at the vinyl group (Reaction 8) and addition of hydrogen radicals to acrylates (Reaction 10) leads to α-alkoxyalkyl radicals. The α-alkoxyalkyl radicals are also able to initiate polymerisation (Reaction 13). Finally, anions form the same type of start radicals as generated by the cationic precursor.

The conclusion can be drawn that after electron-irradiation of acrylates both cations and anions are important precursors of radicals which are able to induce polymerisation and cross-linking.

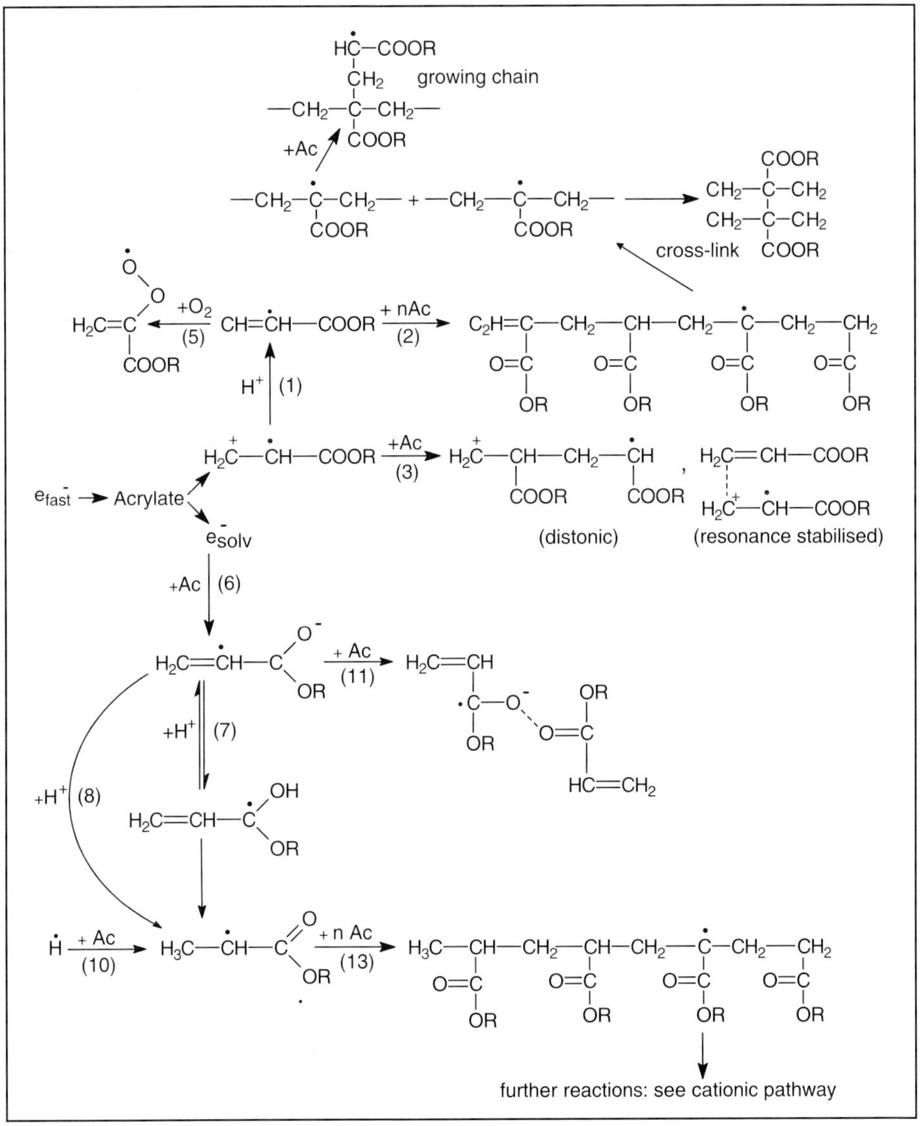

Figure 14 Radical Formation in Electron-Irradiated Acrylates

Less is known about the role of excited acrylate molecules as precursors of radicals. After electron-irradiation excited acrylate molecules are produced either directly or by ion recombination. Their contribution to radical formation is believed to be much less important than that of cations and anions. This is concluded, e.g. from the strong effect of cation and anion scavengers on the electron-induced acrylate polymerisation.

7. Initiation of electron beam curing: Cationic polymerisation

Although radical cations are formed in electron-irradiated vinyl ethers and epoxides, efficient cationic polymerisation is not observed. As pulse radiolysis and electron spin resonance experiments have shown, the primary radical cations are rapidly transformed to radicals, whereas the solvated electron is relatively stable in vinyl ethers and some epoxides. The radical species formed upon electron irradiation cannot induce cationic polymerisation. However, the addition of iodonium, sulphonium or sulphoxonium salts have been used to induce cationic polymerisation under electron irradiation [20].

Reduction of the salt cation by solvated electrons is believed to be a possible source of acid generation. Once the corresponding Brönsted acid is produced, cationic polymerisation proceeds in a way known from photoinitiation.

IV. UV (EB) CURING IN COMPARISON TO THERMAL CURING AND CONVENTIONAL DRYING

After the application of coatings, printing inks and adhesives, an additional process step is needed to transform the liquid film into a solids layer. For solvent based systems traditional operations such as heating by hot air or infrared absorption are used to remove the solvent and to deposit the solid component of the coating, ink or adhesive onto a substrate. If the solid component remains chemically unchanged, this procedure is termed conventional drying (see Figure 15).

Sometimes heat is applied to initiate thermal curing or cross-linking. In most cases additives are used which decay into radicals upon heating. These radicals act as initiators of curing or cross-linking. Thermal curing of powder coatings can be regarded as a special case. The powder is melted by heating and thermally cured at elevated temperatures.

In UV and EB curing the transformation of monomer/oligomer systems is induced by high energy photons or electrons. The curing process leads to the formation of a dense network, which, in comparison to the liquid, shows some volume shrinkage. This shrinkage can be the reason for reduced adhesion on the substrate. On the other hand, electrons penetrating into the substrate are able to generate radicals. Recombination of substrate radicals with radical sites from the network leads to improved adhesion by grafting at the coating/substrate interface.

The principles of curing and conventional drying are schematically shown in Figure 15.

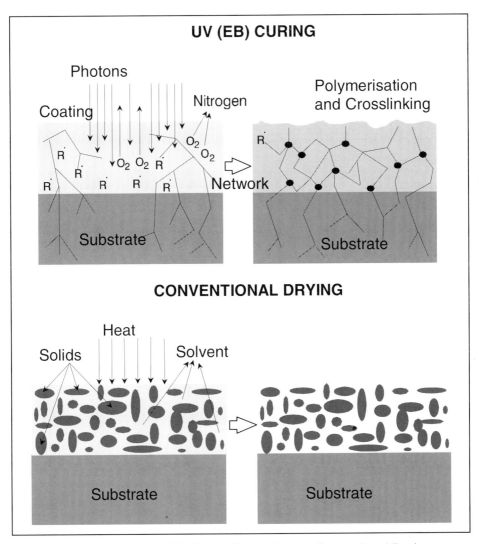

Figure 15 UV(EB) Curing in Comparison to Conventional Drying

Some benefits and drawbacks of UV (EB) curing in comparison with solvent based thermal curing or drying are listed in Table V. In a simplified general overview the comparison is made under commercial, technical, and environmental as well as health and safety aspects.

Table V Comparison of UV, EB and Thermal Curing or Drying

Parameter	UV	EB	Thermal
Commercial			
Capital Cost	+	0	-
Operational Cost	+	0	-
Formulation Cost	-	-	+
Floor Space	+	0	-
Cure Speed	+	+	-
Skill Level Required	0	-	+
Environmental			
No Solvent Release	+	+	-
Energy Consumption	+	+	-
Technical			
Chemical Resistance	+	+	-
Formulation Variety	0	+	+
Curing of Pigmented Films	-	+	+
No Substrate Damage	+	-	0
Low Cure Temperature	0	+	-
Sensitivity to Oxygen	+	-	+
Health & Safety			
Fire Hazard	+	+	-
Radiation Hazard	0	-	+
Irritant Raw Materials	-	-	+

+ Advantage   - Drawback   0 Intermediate

Arguments in favour of the replacement of thermal curing and drying technologies by UV(EB) are mainly lower capital and running costs, lower floor space requirements, higher running speeds, less substrate heating, the high quality of the cured coating or ink, no solvent release during curing and the appearance of less or no skin irritant raw materials on the market.

From the survey given in Table V the conclusion can be also drawn, that unlike the important benefits offered in particular by UV curing, thermal drying still holds a strong position due to advantages in formulation costs and variety, the avoidance of radiation and the lower skill level required.

In some applications EB curing competes with UV curing. However, in all UV(EB) curing applications EB only reaches a market share of less than 10%. Curing of thick pigmented films where high durability and excellent weather resistance are required is still a domain of electron beam applications.

V. ECONOMIC AND ECOLOGICAL DRIVERS FOR THE GROWTH OF RADIATION CURING

In addition to the growth of radiation curing stimulated by the general advantages discussed above, even greater market success may be achieved as the result of technological innovations in curing chemistry and equipment. Certain economic and ecological factors encourage the continuous growth of radiation curing technology. Economic and ecological benefits are offered by recent developments in the chemistry of radiation curable raw materials and in UV&EB curing equipment. To mention only a few:

- Raw materials have been developed, which contain a low amount of volatiles, are less or not skin irritant and increase the range of formulation variety.
- Low-viscosity monomer-free oligomers and water-reducible oligomers close the gap to spray coating applications.
- Oligomers which adhere well to critical substrates open up new applications in metal and glass coating.
- Weathering resistant products are available for outdoor applications.
- More reactive photoinitiators allow lower photoinitiator contents in the formulations or the application of less powerful UV sources.
- Photoinitiator-free UV-curable systems appear on the market.
- Additionally to improved medium pressure mercury lamps new monochromatic UV sources were introduced.
- Low voltage, low cost EB processors are offered.

Technological progress is complemented by progress in the "approval" of acrylates. The American Environmental Protection Agency (EPA) has cancelled the "significant new use rule" for acrylates.

On the coating market radiation-curables are competing not only with solvent based systems but also with high solids, water-based and powder coatings. Figure 16 demonstrates that the relative importance of solvent-based coating is declining in favour of the other, more environmentally friendly, coating systems. In all coating application radiation-curables have the lowest market share but the highest growth rate.

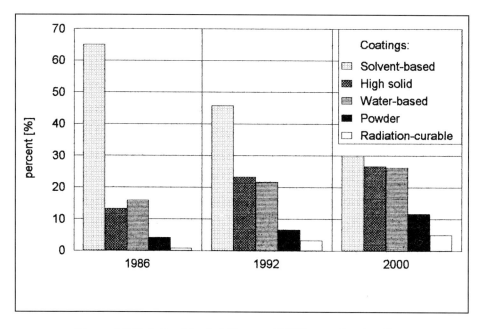

Figure 16 Relative Market Shares of Different Coating Systems

VI. REFERENCES

1. S.P. Pappas, Radiation Curing, Plenum Press, N.Y. (1992)
2. J.P. Fouassier,J.F. Rabek (eds.), Radiation Curing in Polymer Science and Technology, Elsevier, London (1993)
3. P.G. Garratt, Strahlenhärtung, C.R. Vincentz, Hannover (19969
4. M.A.J. Farhataziz, Radiation Chemistry, VCH Weinheim (1990)
5. Ch.E. Hoyle, Radiation Curing of Polymeric Materials, ACS Washington (1990)
6. Oldring, P.K.T. (ed.): Chemistry and Technology of UV and EB Formulation for Coatings, Inks and Paints, Vol. 1–5, SITA Technoloy Ltd., London (1991)
7. H.-G. Elias: An Introduction to Polymer Science, VCH Weinheim, New York, Basel, Cambridge, Tokyo (1997)
8. C. Decker, K. Moussa, Makromol. Chem. 191, p. 963 (1990)
9. M. Wittig, Th. Gohmann, Procceedings RadTech Europe ´93, p. 533 (1993)
10. J.V. Crivello, Adv. Polm. Science 62 ,1 (1984)
11. G. Noren, A. Tortello, J. Vandeberg, Proceedings RadTech North America `90, Vol. 2, p.201 (1990)
12. S. Jönsson,W. Schaffer, P.-E. Sundell, M. Shimose, J. Owens, Ch.E. Hoyle, Procceedings RadTech North America `94, Vol.1, p.194 (1994)
13. J.D. Frank, M. Cekic, C.H. Wood, Procceedings of 32nd Microwave Power Symposium, Ottawa, Canada, p.60, (1997)
14. B. Eliasson, U. Kogelschatz, Appl. Phys. B46, p. 299 (1988)
15. T. Tabata, R. Ito, Nucl. Sci. Eng. 53 , p. 226 (1976)
16. K. Dietliker in Chemistry and Technology of UV and EB Formulation for Coatings, Inks and Paints, ed. P.K.T. Oldring, Vol. 3, SITA Technology Ltd., London (1991)
17. J.P. Fouassier: Photoinitiation, Photopolymerization , and Photocuring, Hanser Publishers, Munich, Vienna, New York (1995)
18. P.J-M. Manus , Procceedings RadTech Europe`89, p. 535 (1989)
19. R. Mehnert, W. Knolle, to be published
20. P. Buijsen, Electron Beam Induced Cationic Polymerization with Onium Salts (Thesis Delft University of Technology), Delft University Press (1996)

CHAPTER II

INDUSTRIAL APPLICATIONS OF RADIATION CURING

I. INTRODUCTION

Radiation curing technology seems to be paralleling the growth of electrical technology in industry as the 19th century came to a close. From the time the electric motor was introduced, people talked about its productivity potential, but it took more than five decades really to learn how to distribute and use electricity efficiently. Diverse applications were generated in the 1920's. Can we today imagine life without electricity? Well, one of the technologies that were generated by electrical technology is radiation curing. We've been talking about and experimenting with ultraviolet (UV) and electron beam (EB) technology for about 30 years. As it was with electrical technology, manufacturers have processes in place using other technologies. In order to grow, radiation curing must replace some of these technologies. We must figure out how to convince these manufacturers that radiation curing technology is superior. We need to sell the benefits of radiation curing to these manufacturers and to their customers. As the 20th century comes to a close, we have successfully demonstrated multiple industrial applications for this technology which offers many benefits, including:

        high quality and performance,
        high productivity,
        ease of use,
        friendliness to the environment, and
        low energy consumption.

In the 1990s, we have seen much growth in the use of radiation curing technology worldwide. Many factors have contributed to this growth:

1. Information dissemination – The worldwide web now provides instant information just by typing in key words in a search engine. We can search for equipment and material information or learn about applications and find contacts in the field in the comfort of our office at work or at home. This important source of information will continue to improve at a fast pace. More suppliers will participate and will include more information on their products. Several trade associations have set up web pages allowing searchers to obtain up-to-date information on events, to link to other sources, and even to initiate or participate in on-line discussions with anyone visiting the site. These pages provide global visibility to the associations and to the technologies they represent. In addition to the education provided by suppliers to their customers, more organizations are offering seminars and conferences covering

various aspects of radiation curing. More trade magazines are featuring radiation curing articles. More books and manuscripts are available on the subject.
2. Materials – an array of new monomers, oligomers, photoinitiators, and additives continue to lead to improved formulations. Some new substrates and composite materials rely on radiation curing or processing. Higher volume consumption, improvements in manufacturing, and the introduction of more competitors have translated to a decrease in the cost of some materials.
3. Equipment – smaller, lower cost, and more reliable EB units; higher intensity, better heat removal, more wavelength specific UV curing systems; multigraphic printing presses; better controlled coaters.
4. Markets – high growth applications have included UV coatings on compact disks, UV flexo printing, EB release coatings, various UV applications on vinyl cards (credit, debit, ATM, I.D., or prepaid telephone cards, so-called 'smart cards'), on business forms, and on optical fibres. New developments in powder coatings which are fused by infrared (IR) and cured by UV or EB widen the range of potential applications, especially for 3-dimensional objects. Declining markets include vinyl flooring (losing ground to the new high pressure laminate flooring), magnetic media binders (losing ground to compact disks and digital disks which also utilize radiation curing technology).

Radiation curing applications are best categorized according to the function of the cured formulation as listed in Table 1.

Table 1 Radiation Curing Applications

1. COATINGS
2. PRINTING INKS
3. ADHESIVES
4. PLASTIC PARTS

II. RADIATION CURABLE COATINGS

Coatings are applied to a surface. They can be functional or decorative or both. The surface or substrate to which the coating is applied, the method of application, the required properties, and end-use conditions all guide the formulator in designing the best formulation possible.

Functional coatings improve the surface by:
- protecting it from abrasion, scratch, mar, chemicals,
- providing different properties such as release, slip, adhesion, electrical conductivity or insulation, antifogging, flame retardance,
- acting as a barrier to various liquids or gasses.

Decorative coatings are applied to:
- change appearance (color, gloss or matt finish, texture),
- hide surface (imperfections, electrical circuitry, etc…).

Coatings are most often classified according to the substrate they are applied to, as listed in Table 2. Radiation curable coatings have been developed for all types of substrates. Some suppliers specialize in coatings for certain substrates.

Table 2 Radiation Curable Coatings

1. PAPER AND PAPERBOARD
2. WOOD
3. PLASTICS
4. METAL
5. GLASS AND CERAMIC
7. MISCELLANEOUS

1. Applications on paper and paperboard

The best known paper and paperboard application for UV and EB coatings is referred to as an overprint varnish (OPV). It can be applied at the end of the printing line or off-line in a separate operation. As implied, the coating is applied over a printed surface to protect the ink layer and to improve the appearance of the product. High gloss varnishes can really make a package stand out on a shelf. For additional appeal, some surface designs include a high gloss varnish with smaller areas of matt finishes or vice versa. This is referred to as spot curing.

Examples of radiation curing applications on paper substrates are listed in Table 3 below.

Most of the EB units sold recently have been for release coating applications, indicating greater growth in this area. New chemistry has been commercialized but further technical developments are needed to satisfy this market segment.

Table 3 Radiation Curable Paper Coating Examples

1. OVERPRINT VARNISHES FOR
   - BOOK, MAGAZINE, BROCHURE COVERS
   - PAPER CARTONS
   - CEREAL, GAME, TOY, COSMETIC, PHARMACEUTICAL BOXES
   - FROZEN FOOD PACKAGES
   - VIDEO TAPE BOXES
   - LIQUID CONTAINERS (JUICE, MILK, WINE)
   - GIFT ENCLOSURES (BAGS, WRAPPING PAPER)
   - LABELS.

2. BASECOATS FOR:
   - PHOTOGRAPHIC, LASER OR INKJET PRINTER PAPERS
   - METALLIZED OR SPUTTERED PAPERS
   - SHEETFED OFFSET PAPERS.

3. HIGH OR CONTROLLED GLOSS COATINGS ON SPECIALTY PAPERS

4. RELEASE COATINGS FOR:
   - TEXTURED CASTING PAPERS
   - ROLL OR SHEET LABELS
   - TAPES.

5. BINDERS FOR:
   - GRINDING MEDIA (ABRASIVE PAPERS)

2. Applications on wood

Manufacturers of wood products have found that the use of radiation curing technology improves quality while significantly cutting turn-around time, labour, and waste. Poor quality wood such as chipboard can be upgraded to furniture quality by EB laminating printed wood grain or other decorative papers and applying a finish layer. 3-dimensional objects such as chairs, sporting goods, musical instruments can be coated and cured in custom designed curing chambers. High quality and durable finishes can be obtained.

Coatings are applied by roll or curtain to flat products. 3-D products are typically sprayed or dipped in fewer operations than with solvent based systems. Waterborne UV coatings continue to make in-roads in spraying applications. UV powder coatings are also finding a niche in 3-D wood products. At the most recent North American RadTech conference in Chicago, a floor demonstration of the application of UV powder coating to pressed wood coasters took place.

Table 4 Examples of Radiation Cured Wood Products

```
1. FLAT PRODUCTS
     - SHELVING
     - CABINETRY
     - PREFINISHED FLOORING

2. 3-DIMENSIONAL PRODUCTS:
     - CHAIRS
     - TABLES
     - GUITARS
     - BROOM HANDLES
     - PICTURE OR BOARD FRAMES
     - MOLDINGS
```

3. Applications on plastics

New plastic materials are constantly being introduced, replacing and outperforming paper, glass, metal, or wood in several industrial applications. Plastics encompass thin flexible films to rigid 3-d parts. One of the major applications of EB technology is in the modification of plastics EB crosslinking strengthens films and modifies their temperature resistance window. Electron beam irradiation grafting of films can change properties such as wettability, flame retardance, diffusion. EB and UV coatings are applied to all sorts of plastic materials for various reasons.

Table 5 Examples for Radiation Curable Coatings on Plastics

1. COATINGS ON HEAT SENSITIVE SUBSTRATES (PVC, PE, PP STYRENE, PET, PC, etc…)
2. BASECOATS FOR:
    - AUTOMOTIVE PARTS
    - METALLIZED PLASTICS
    - VINYL FLOORING (TILES, SHEET)
3. TOPCOATS FOR:
    - COMPACT DISKS
    - VINYL (IMITATION LEATHER, VENEER FILMS, CARDS, FLOORING, UPHOLSTERY, WALL COVERINGS, NOTEBOOK COVERS, etc…)
    - CONTAINERS (CUPS, BOTTLES, TUBES)
4. HARDCOATS FOR:
    - RIGID PLASTICS (PC/MMA WINDOW GLAZINGS, WINDSHIELDS, GOGGLES, SPECTACLE LENSES, etc…)
    - AUTOMOTIVE HEADLIGHTS, BUMPERS,
5. BINDERS FOR:
    - MAGNETIC MEDIA (TAPES, DISKS)
6. RELEASE FILMS (LABELS, MEDICAL PRODUCTS, etc…)

4. Applications on metal

Formulations with excellent adhesion to metal at competitive cost have opened up this market, especially for can coatings. Examples of radiation cured metal products are listed in Table 6. It is still a challenge to obtain flawless overprint coatings on metal. Balancing the flow additives and a minimum dwell is required to allow the coating to flow prior to curing. The smooth finish and highly crosslinked nature of UV formulations make for an excellent erasable board surface. Formulations designed for exterior durability provide excellent protection to solar reflectors and galvanized steel pipe. Another important application on metal is photoengraving. UV coating is applied to the metal substrate as a mask to resist etching. Once the etching process is done, the UV coating is stripped away.

Table 6 Examples of Radiation Cured Metal Products

1. OVERPRINT COATINGS ON:
    - NAMEPLATES
    - DECALS
    - I.D. LABELS
    - FOIL
    - CANS
    - TUBES
    - SIGNS
    - PAILS
2. BASECOATS FOR:
    - METALLIZING (CHROME PLATING, SPUTTERING)
3. COATED COILS
4. SOLAR REFLECTORS
5. ERASABLE BOARDS
6. GALVANIZED STEEL PIPE
7. ETCH RESIST FOR PHOTOENGRAVED NAMEPLATES

5. Applications on glass and ceramic

Optical fibre and ribbon cable applications will continue to grow for several years in both the telecommunications and automotive industries. UV coatings may not be as durable, but they offer a better work environment, lower energy consumption, and safer product than the heavy metal glazes that are fired at extremely elevated temperatures.

Table 7 Examples of Radiation Cured Glass or Ceramic Products

1. COATINGS ON OPTICAL FIBRES AND RIBBON CABLES
2. PROTECTIVE COATINGS ON BACK SIDE OF MIRRORS
3. WINDOW GLASS COATINGS
4. DECORATIVE COATINGS ON GLASS SCULPTURES
5. GLASS BOTTLE COATINGS
6. IMITATION ETCH GLASS COATINGS
7. CERAMIC OR GLASS CUP/MUG COATING FOR SUBLIMABLE/TRANSFER DYE APPLICATION AND FIRED GLAZE REPLACEMENT.

6. **Miscellaneous coating applications**

Table 8 Examples of Radiation Cured Products on Miscellaneous Substrates

1. LEATHER COATINGS
2. TEXTILE/FIBRE COATINGS
3. COMPOSITE COATINGS

III. RADIATION CURABLE INKS

Inks are highly pigmented formulations which are used to print an image onto a substrate. There are various printing methods and inks are classified accordingly. Ink application properties such as consistency, viscosity, pigment level, are dictated by the printing method. The paste type inks (offset, letterpress, screen, pad) were the first radiation curing applications because it was easier to formulate high viscosity inks with the available raw materials. Since the availability of lower viscosity inks, there has been explosive growth in liquid inks (flexo, gravure, ink jet). Printers have embraced radiation curable inks because of their ease of use, stability on press, instant cure, clean color, heat and rub resistance. There is a growing trend toward multigraphic printing, that is, having different printing methods on one line for the ultimate in flexibility. It allows the printer to bid for a range of print jobs to run on one line. In the new designs, the printing stations are interchangeable so that the printer can install whatever printing method is desired for that station. A significant lowering in ink costs has made it easier to justify UV/EB inks.

Table 9 Examples of radiation curable inks

1. INTAGLIO
2. OFFSET / LITHO
3. SCREEN
4. LETTERPRESS
5. FLEXO
6. GRAVURE
7. TRANSFER PAD PRINTING
8. HOT STAMP
9. JET PRINTING

1. **Intaglio Inks.** Intaglio inks have high viscosity and are typically applied to security papers and bank notes. They provide a raised image. Because of the ink layer thickness and desired opacity, EB inks have been preferred over UV inks.
2. **Offset / Litho Inks.** Offset or litho inks are paste inks used in offset or litho printing. UV / EB curable inks are replacing sheetfed offset oxidative inks, heatset litho inks, and dry offset inks on many lines. EB is chosen over UV for some food packaging (indirect contact) applications because lower extractables and lower odour can be more easily and consistently obtained than with UV. These applications include: microwavable meal packages, juice containers, cereal boxes.

UV is more common than EB in all other roll, sheetfed, and dry offset applications because of its lower capital cost. Roll applications that are sheeted after cure (in-line or off-line) and sheetfed applications include: paperback book and magazine covers, hard cover book sleeves, labels, business forms, brochures, etc. Dry offset printing is mostly used on containers (bottles, cups, bowls, pails of all sizes), covers, and caps.

3. **Screen Inks.** UV technology has made screenprinting a much more efficient printing process. The inks stay wet on the screen for easy application and low maintenance, but they instantly cure when exposed to UV radiation. No more racking of wet prints. UV ink odour is reported to be much less offensive than that of solvent based inks. Products printed with UV inks are sold on appearance and durability. Some of the products printed with UV inks include: vinyl awnings, vinyl decals, flags, banners, signs, presentation folders, labels.

4. **Letterpress Inks.** More letterpress units have been replaced by less expensive flexo units, but letterpress is still used for some lower quality label printing and UV letterpress inks are preferred.

5. **Flexo Inks.** Most ink suppliers now offer UV flexo inks. With all the experience gained in UV flexo printing, it can no longer qualify as a new technology. Due to the relatively low cost of flexo equipment, this application has grown in leaps and bounds worldwide. Some companies combine UV (interstation) and EB (terminal) for optimum cure.

6. **Gravure Inks.** Now that low enough viscosity UV/EB inks are available, it is up to the gravure printers to make a commitment to this technology so that the inks can be optimized for this application. It may be that EB is more likely to meet the cure speed requirements.

7. **Transfer Pad Printing Ink.** Transfer pad printing is a unique process which allows printing on curved or odd-shaped, textured, raised or recessed parts. Printing of logos or promotions on a golf ball is an excellent example of the flexibility of this printing method. It prints uniformly on the curved object's recessed and raised areas. Transfer pad printing finds application on all sorts of products and substrates in several industries. In addition to manufacturer identification of products, it is used to print warranty information, part numbers, etc…

PAD PRINTING INKS

UV	vs.	EPOXY
♦ single component		♦ 2 components (must be mixed just prior to use)
♦ leftover ink is reusable		♦ unused ink must be cleaned out and thrown away
♦ consistent colour		♦ colour is a function of mixing procedure

- instant cure, no drying racks, fast turnaround
- must be racked (several hours) until fully cured

8. **Hot Stamp Inks.** These inks are transferred with heat and pressure to the substrate from a carrier film, typically polyester. UV inks are more durable.

9. **Jet Printing Inks.** The bulk of solvent or water based jet printing ink systems consists of viscosity adjustment with solvent. With UV jet inks, viscosity is stable because there is no evaporation taking place. This also means that these inks do not clog the jets, another major maintenance headache.

IV. RADIATION CURABLE ADHESIVES

Adhesives are classified according to their function: laminating, pressure sensitive, heat seal, structural, sealant. They represent totally different markets.

1. **Laminating Adhesives.** For UV curing of a laminating adhesive, one substrate must be transparent to UV. For EB curing, penetration depth is a function of material density, not colour or type. Radiation cured laminating adhesives typically offer high bond strengths and improved chemical and heat resistance. Examples of industries that use UV or EB laminating adhesives are: packaging (for multilayer bags or boxes; bonding of different substrates for different functions such as bonding polyester to heatsealable polyethylene), construction (bonding printed paper, wood veneer, or film to particleboard, pressboard or plywood; laminating solar films to glass windows, or glass to glass for reduced weight; electronics (applying conductive adhesives between two substrates for shielding; making multilayer video screens); automotive (glass to glass rear windows, flexible films to rigid plastics or metals for interior parts,); shoe industry (laminating fabric to foam or rubber, foam to rubber, foam to foam); greeting cards or gift wrapping (laminate foil to paper).
2. **Pressure Sensitive Adhesives.** Although some unique UV curable pressure sensitive adhesives are in commercial use, the common pressure sensitive adhesive has not been replaced. It is still a performance issue rather than cost. Pressure sensitive adhesives tend to have a narrow cure window for optimum performance.
3. **Heat Seal Adhesive.** At first the idea of radiation curable heat seal adhesives may sound impossible since radiation curing is synonymous with crosslinking which increases heat resistance. It is however feasible to transform a liquid to a solid with radiation and still obtain a heat sealable adhesive. As printers become accustomed to their UV/EB curing lines, they are interested in streamlining operations by doing as much on line as possible. Adding a print station for the application of EB or UV curable heat seal adhesive for closing packages increases efficiency.

4. **Structural Adhesives**. UV is used to cure structural adhesives for the electronics (component tacking), consumer goods (wine glass stem to bowl), medical devices (syringes, scopes, optical fibre bonding), and automotive industries (metal or plastic rear view mirror stem to glass windshield).The best feature of UV structural adhesives is that they have an infinite pot or work life but cure instantly upon exposure to UV. Because the area of adhesive is typically small, this type of curing is referred to as spot curing. Small lamps which are typically operated with a pedal are used for curing. Combining UV with other cure mechanisms has helped overcome shadow curing. Some opaque parts can be fastened with dual cure adhesives. For example, when adhesive is applied between two opaque parts, the exposed adhesive on the perimeter can be rapidly cured with UV and provide enough bond strength until the hidden adhesive cures by an anaerobic or other secondary mechanism.
5. **Sealants**. UV curable formulations are used to seal can ends in 3-piece cans. Other applications include sealing holes or pores for a smoother finish. UV sealants are also used to repair leather or vinyls. Many windshield repair kits contain UV curable sealants and a portable long wave UV lamp.

V. RADIATION CURING FOR MANUFACTURING PLASTIC PARTS

UV formulations can be formed into products and cured instantly. Laser beams are used in conjunction with computers to generate photopolymer printing plates or to fabricate 3-dimensional complex shapes one layer at a time. The latter process is called rapid prototyping. The technology is also used for photoimaging body parts in 3-dimensions prior to surgery.

Castings of UV formulations can be made and cured in place connecting other components as opposed to fastening or gluing other materials.

Many soft and disposable contact lenses are cast UV materials.

High viscosity UV or EB formulations can easily be extruded and cured close to the extruder die. Products of various shapes can be generated easily.

CHAPTER III

UV CURING EQUIPMENT - POLYCHROMATIC UV LAMPS

I. POLYCHROMATIC UV RADIATION FOR CURING

To induce UV curing, light must be absorbed by a suitable photoinitiator which is part of the radiation-curable formulation. In the curing reaction of a given chemical formulation, choice of the light source critically determines the rate of chemical conversion. Some basic characteristics which are of importance for industrial UV curing light sources are summarised as follows:

- The photon energy needed is determined by the absorption spectrum of the photoinitiator. The emission spectrum of the light source should match the absorption spectrum of the photoinitiator as closely as possible or, in more general terms, the spectrum of the reactant producing the curing-initiating species.
- The rate of photons emitted by the light source should be high enough to induce fast chemical conversion. As shown in Chapter I, Figure 10, the polymerisation rate of a radical system is proportional to the square root of the number of photons absorbed per second. Of special importance is the radiant power of a light source in the UV range. Thus, a high photon rate and a matched emission of the light source are necessary conditions to ensure fast conversion.
- The geometry of the light source should allow homogeneous illumination of the coating to be cured. For most curing applications, a linear cylindrical geometry of the light source is a useful choice. As complete curing requires a certain dose, an optimised light source and reflector geometry is the third important condition for fast conversion. For UV curing applications, a linear cylindrical (tubular) geometry of the light source and a specially shaped aluminium reflector are often used. Light source and reflector geometry together with UV radiant power determine peak irradiance at the coating surface.
- Infrared emission from the light source should be kept at a tolerable level. Substrate and coating might be sensitive to heat. Thus, undesired heat transfer can set limits for the application of a distinct light source.

Emission generated from a dense mercury vapour plasma fulfils all basic requirements of a UV curing light source: there are many strong emission lines and some continuous background radiation in the UV-VIS range. The UV radiant power amounts to up to 35 % of the total radiant power, and up to 10^{20} UV photons per second can be emitted per cm plasma column. Linear cylindrical plasma columns with a length of 2m or more are technically viable, and various measures can be taken to decrease the heat transfer to coating and substrate.

II. LIGHT EMISSION FROM A DENSE MERCURY VAPOUR PLASMA

The term plasma describes an ionised gas, consisting of a mixture of interacting positive ions, electrons, neutral atoms or molecules in the ground state or any higher state of any form of excitation, as well as of photons. Because charge carrier pairs are generated by ionisation, the plasma as a whole remains electrically neutral [1].

Plasma is formed by energy transfer to a system of atoms or molecules until the gas phase and a high degree of ionisation are obtained. In the case of polychromatic UV curing light sources, the plasma is generated using mercury vapour, and the source of energy is electricity. Mercury vapour gas discharges excited either by AC (DC) current, radio frequency or microwave absorption, form the physical basis for an important class of light sources used for curing. Mercury is quite a unique fill gas for such discharges: it has a fairly low ionisation potential and sufficient vapour pressure in the temperature range used, it is chemically inert towards the electrode and wall materials and has a favourable emission spectrum. At the same time, the excited levels of the mercury atom are high enough to allow excitation transfer to other metals such as iron, gallium, lead, cobalt etc., thus enabling modification of the spectral output of the light source.

1. Light emission from a mercury gas discharge

Let us consider a non-conducting tube filled with argon of a few mbar and a certain amount of (liquid) mercury. If electrodes are placed at each end of the tube, and a DC voltage is applied, a current starts to flow through the gas. With increasing current, the voltage across the electrodes shows a profile characteristic for a gas discharge [2,3,4].

A typical voltage-current characteristic of a gas discharge is shown in Figure 1. For the design of intensive light sources, only self-sustaining discharges such as glow and arc discharge are of importance.

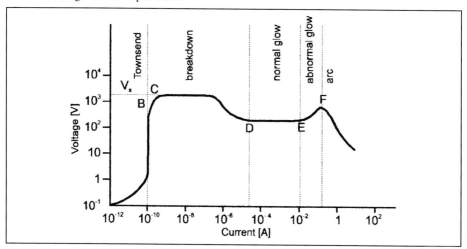

Figure 1 Voltage-Current Characteristic of a Gas Discharge

In a glow discharge, the number of mobile positive ions striking the cathode is sufficient to cause enough electron ejection for a steady state, self-sustaining discharge. However, photon emission is insufficient in a normal glow discharge, because the mercury vapour pressure is in the range of less than 10^{-3} mbar, and the resulting population density of mercury excited states is low. If the voltage is further increased, more positive ions strike the cathode thereby increasing electron emission and cathode temperature. In the abnormal glow discharge region, the mercury vapour pressure rises to about 10^{-2} mbar. Effective population of the lowest electronically excited mercury levels takes place, resulting in the emission of strong 185 and 254 nm radiation, which can be used for technical applications. The low pressure mercury lamp operates in the region of the abnormal glow discharge.

When the bombardment of the cathode by positive ions induces thermionic electron emission, discharge current, gas temperature and mercury vapour pressure increase by several orders of magnitude. A typical arc discharge is obtained at mercury vapour pressures from 0.1 to 10 bar and currents up to 10 A. The higher electronically excited mercury levels are populated as well, and even ionisation takes place. Medium pressure mercury lamps operate in the region of arc discharge. Various emission lines and some background radiation appear in the ultraviolet and visible ranges of the spectrum. The most intensive transition is at 365 nm. However, 185 and 254 nm emission is suppressed by strong self-absorption through mercury ground state atoms.

One important peculiarity of arc discharge is its negative voltage-current characteristic. In a stable arc discharge, a balance is established between the formation of new charge carriers and the loss of electrons and ions by recombination. If the balance is disturbed, e.g. by increasing the current, an increase in charge carriers takes place which will not be compensated for by a corresponding increase in recombination. The internal impedance of the discharge will drop. That leaves an excess voltage across the electrodes resulting in further current increase, thereby creating a catastrophic cascade effect. To avoid lamp destruction immediately after ignition, a suitable impedance has to be placed in series with the lamp. For common AC driven mercury arc lamps, a combination of inductive and capacitive load is used to compensate for the negative characteristic of the arc.

To describe the mechanism of photon generation from mercury gas discharges, collisional and electrostatic interactions beween the electrons, ions and atoms of the mercury vapour plasma which can lead to the population of mercury atomic states, have to be considered first. Secondly, the term scheme (Grotrian or energy transition diagram) of the mercury atom is required, summarising the energetic positions of excited states and indicating possible optical transitions.

Electrons accelerated in the electric field of the gas discharge, can either interact with ground state gas atoms by elastic collisons or by electrostatic (Coulomb) forces exercised to outer shell electrons of the mercury atoms. Because of the large mass differences of the collisional partners, very little energy is exchanged by collisions between electrons and atoms. Electrostatic interaction leading to excitation and ionisation of the mercury atoms is the basic phenomenon underlying photon emission.

At low pressures, e.g. in the region of glow discharge, the frequency of electron-atom collisions is low, and consequently the mean free path length of electrons is high.

While ions lose their entire energy by collisions, electrons can achieve a high kinetic energy. The energy distributions of electrons, and of atoms and ions vary considerably. If the energy distribution is characterised by a temperature, the "electron temperature" is much higher than that of atoms or ions. In the mercury glow discharge, the electron temperature is of the order of 10^4 K, while atom and ion temperatures are in the region between 3×10^2 and 10^3 K.

At a higher pressure the collision frequency is much greater, the kinetic energy gained by the electrons decreases considerably, but also the frequency of energy transfers increases dramatically. The result is that a local thermal equilibrium is achieved. Electrons, atoms and ions are at (nearly) the same temperature. This is especially the case in the arc discharge of a medium pressure mercury lamp.

Figure 2 illustrates the effect of the gas pressure on electron, atom and ion temperatures in a mercury gas discharge. The plasma temperature is in the region of 6000 K corresponding to an energy kT of 0.54 eV.

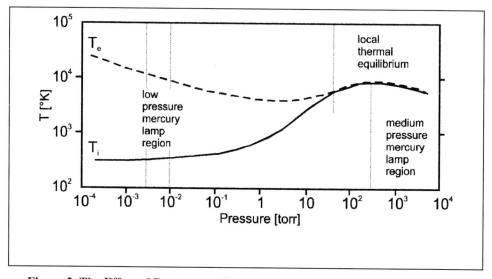

Figure 2 The Effect of Pressure on Electron (T_e) and Ion Temperature (T_i) in a Mercury Gas Discharge

To understand the basic characteristics of photon emission from an atom, the atomic energy level and transition diagram can be favourably used. Figure 3 shows this diagram for mercury excited in an arc discharge at medium pressure [5].

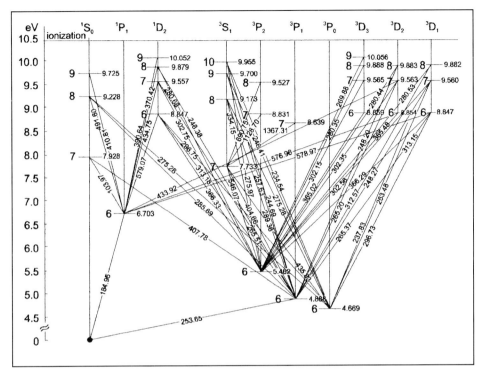

Figure 3 Grotrian Diagram for Atomic Mercury

The emission of a photon by an atom results in a change of the electronic state of the atom. The energy loss is equal to the difference in the energy of the upper (E_h) and lower (E_l) energy level.

$$E_h - E_l = h\nu, \qquad (1)$$

where h = Planck's constant and ν= frequency of the photon emitted.

According to quantum mechanics, the energy of a given electronic state is determined by the quantum numbers combined in a term symbol $n^{(2S+1)}L_J$. In Figure 3, the energy levels are denoted by the principal quantum number n (6 - 10), the spin quantum number S, expressed as singlet (1) or triplet (3) "multiplicity" (2S+1), the total orbital angular momentum L (expressed as S,P and D) and the total angular momentum J L and S combine as vectors to form a resultant J (0,1,2,3). However, not all possible energetic transitions are permitted. There are selection rules for radiative transitions. For mercury having two electrons in the outer shell, the permitted radiative transitions are those with $\Delta S = 0$, $\Delta L = 0$ or ± 1 and $\Delta J = 0$ or ± 1, except for $J = 0 \rightarrow 0$. There are no restrictions on changes in the principal quantum number n.

The transitions indicated in Figure 3 lead to discrete lines in the mercury spectrum. Superposition of lines, which become broader at increasing pressure, and a continuum attributed to radiation from the recombination of mercury cations and electrons, finally lead to the typical spectral output of a medium pressure mercury arc lamp (see Figure 4).

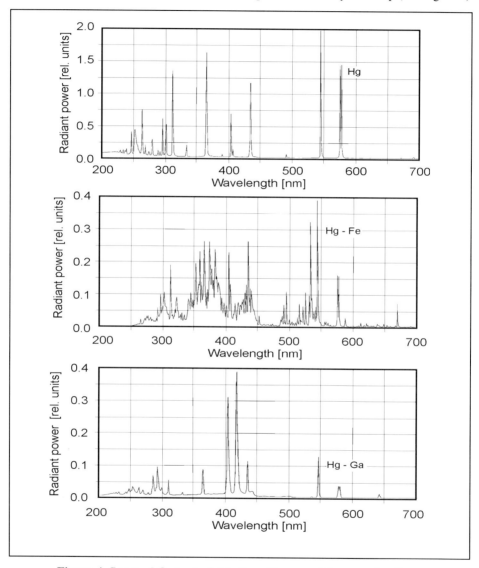

Figure 4 Spectral Output of Medium Pressure Mercury Arc Lamps

Although the spectral output of a mercury arc discharge matches the absorption spectra of many photoinitiators used for curing very well, there are regions where more radiant power would be desired for special curing application. This is of particular importance when pigmented systems have to be cured, where pigment is strongly absorbing at wavelengths below 360 nm.

In such cases metal halides can be added to the mercury discharge to generate transitions where enhancement is required. Gallium and iron halides are often used in this case. Gallium addition produces strong spectral lines at 403 and 418 nm, whereas iron addition leads to strong multiline emission between 350 and 400 nm (see Figures 4 and 5).

Ultraviolet light showing wavelengths from 40 to 400 nm is part of the electromagnetic spectrum.

Ultraviolet radiation is divided into:
- Vacuum UV : 40-200 nm
- UV C : 200-280 nm
- UV B : 280-315 nm
- UV A : 315-400 nm.

Vacuum UV radiation is strongly absorbed by quartz which is used as envelope material for lamps. For this reason, and also because of its small penetration depth, VUV radiation is not considered for curing.

Some of common units and definitions used with UV radiation are briefly summarised in Table I.

Table I Units and Definitions Relating to UV Light Sources

Wavelength		Wavenumber	
Nanometre (nm)	10^{-9} m	1/Wavelength, $1/\lambda$	cm^{-1}
Micrometre (μm)	10^{-6} m		
Energy		**Frequency**	
J (Joule)		Hertz (s^{-1})	
Energy of a photon $E = h\nu = hc/\lambda$			
$= 1.986 \times 10^{-16}/\lambda$		**Power**	
(wavelength λ in units of nm)		1 Watt = 1 J s^{-1}	
		$= 5.034 \times 10^{15}/\lambda$ photons s^{-1}	
h is Planck's constant 6.626×10^{-34} Js		(λ in units of nm)	
c is velocity of light 2.998×10^{8} ms^{-1}			

2. Light emission from a microwave excited discharge

Microwave radiation, such as that generated from a magnetron, can be directly transmitted into a quartz bulb containing argon as a start gas and a certain amount of liquid mercury. The high electric field strength produced by the microwave inside the bulb ionises the argon and heats it up, vaporising the mercury. Mercury becomes excited and to some extent ionised. As a result, the gas fill is acting as an effective microwave energy sink. A mercury vapour plasma is formed which can be operated up to mercury pressures of 20 bar. The technical application of microwave-excited medium pressure mercury plasma light sources was pioneered by UV Fusion Systems Inc. Gaithersburgh (USA) [6,7,8] and has found wide use.

From a microwave-powered mercury vapour plasma intense photon emission is observed. As the plasma conditions such as electron, atom and ion temperatures as well as mercury pressure are similar to that of a medium pressure mercury arc discharge, the spectral output of microwave-excited medium pressure mercury lamps is comparable to that of the corresponding current-powered arc lamps (see Figure 5). In the same way as mentioned for mercury arc discharges, metal halide addition is also frequently used in microwave-powered lamps to enhance the radiant power in the UV and visible range.

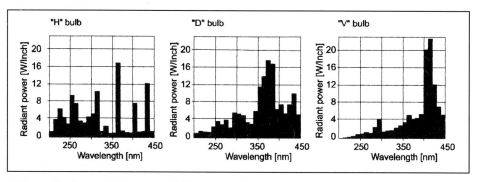

Figure 5   Spectral Output of Microwave -Powered Electrodeless Medium Pressure Mercury Lamps

III. POLYCHROMATIC LIGHT SOURCES FOR UV CURING

1.   Low pressure mercury lamp

Low pressure mercury lamps consist of a melted quartz (vitreous silica) envelope of tubular, spiral or U-shaped form with electrodes at both ends. Two basic types of low pressure mercury lamps are distinguished depending on the mechanism of electron ejection from the cathode: cold or thermionic (hot) cathode lamps. In the latter case a tungsten wire heated oxid cathode is used as electron emitter. The gas filling is a mixture of argon and mercury. A few mbar of argon are needed to ignite the discharge. The mercury pressure is kept in the range of 10^{-2} to 10^{-3} mbar and corresponds to the mercury vapour pressure at the operating temperature of about 40°C. Cold cathode lamps operate in the region of abnormal glow, whereas hot cathode lamps show the characteristics of an arc discharge. Because of the negative voltage-current characteristics of the arc discharge, a hot cathode low pressure mercury lamp has to be operated with ballast. In most cases low pressure mercury lamps are AC-operated from mains, and the ballast consists of an impedance load placed in series with the lamp (see Figure 6).

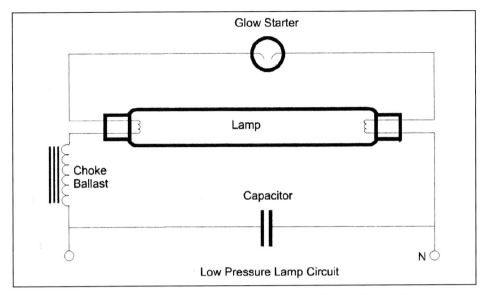

Figure 6 Hot Cathode Low Pressure Mercury Lamp: Lamp Circuit

However, the arc discharge current has to be kept in a range where only the lowest energy levels of the mercury atom are excited. In this case almost all radiation emitted by the low pressure mercury lamp is resonance radiation at 253.7 and 185 nm (see the Grotrian diagram given in Figure 2). Figure 7 shows the spectral output of a typical low pressure mercury lamp. About 80% of the total radiant power is emitted by the 253.7 nm line. Application of special Suprasil quartz as an envelope also allows a utilisation of 185 nm radiation if necessary.

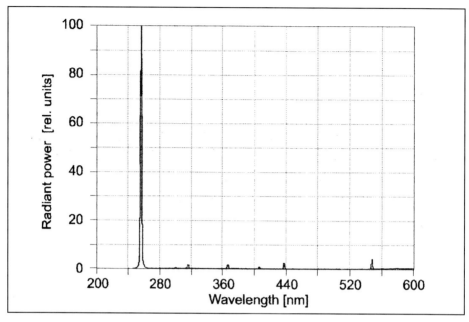

Figure 7 Spectral Output of a Low Pressure Mercury Lamp

As shown in Table I, the total radiant power of low pressure mercury lamps can reach up to 40% of the electrical power.

Low pressure mercury lamps operate slightly above the ambient temperature. In most cases cooling is not required. However, the low output radiant power of <1 W cm^{-1} and the low penetration depth of the high energy photons emitted, set limits for curing applications of low pressure mercury lamps.

Table II Characteristics of Some Low Pressure Mercury Lamps

Type	NN 15/44	NN 50/97	NNQ 8/18	NNI 200/10	NIQ 120/80	UCL-MM 10-26
Manufacturers	Hanau	Hanau	Hanau	Hanau	Hanau	ABB
Electric power [W]	15	50	10	200	120	1720
Total radiant power [W]	6	18	4	55	38	470
Spectral range [nm]	254	254	185-254	254	185-254	
Temperature range [°C]	0-30	0-30	0-30	0-70	0-70	
Lamp voltage	58	125	42	105	95	
Lamp current [A]	0.36	0.54	0.36	2.5	1.8	
Radiative length [mm]	350	920	152	1060	800	1700

It is only the hot cathode lamp which is used in special curing applications, such as surface pre-curing and final curing as well as physical matting.

It seems to be likely, however, that cold monochromatic excimer light sources such as Xe_2 (172 nm) and KrCl (222 nm) excimer lamps, showing similar spectral output but higher radiant powers, can easily replace low pressure mercury lamps in curing applications.

2. Medium pressure mercury arc lamp

(i) Construction

It is estimated that in over 95% of UV curing applications medium pressure mercury lamps, either arc or microwave-powered, are used. This is not surprising, because medium pressure mercury lamps, in a nearly ideal manner, meet all the requirements of a curing light source as defined at the beginning of this chapter.

The efficiency of converting electrical into UV/VIS radiant power amounts to 50%. Up to 30% of the total spectral output is produced in the wavelength range from 200 to 400 nm. Lamps with power levels from 40 to 250 Wcm^{-1} are commercially available. In addition, the linear cylindrical medium pressure mercury arc lamp can be made in any length from centimetres up to more than 2 metres.

The construction of a typical tubular medium pressure mercury arc lamp is schematically shown in Figure 8.

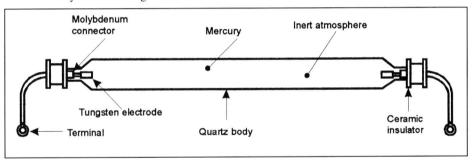

Figure 8 Medium Pressure Mercury Arc Lamp

The lamp consists of a vitreous silica tube of 18-30 mm diameter and 1-2 mm wall thickness. Vitreous silica is used because of its very low temperature expansion coefficient and its excellent transparency in the UV range. The widest and most uniform spectral transmission profile is obtained for synthetic vitreous silica, which is known by the tradenames Suprasil and Suprasil W (see Figure 9).

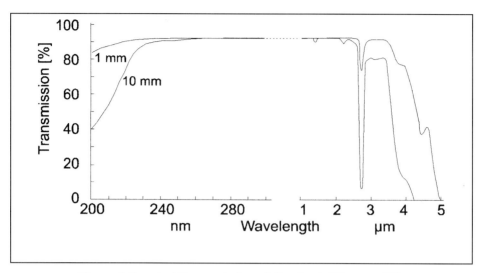

Figure 9 Spectral Transmission of Synthetic Vitreous Silica

Other types of vitreous silica show transmission losses below 240 nm and strong absorption around 2700 nm in the infrared. They are less expensive than synthetic silicas and therefore commonly used for lamp manufacture. The electrodes of a mercury arc lamp are compact thoriated tungsten rods, wrapped by a tungsten wire double coil carry alkaline earth oxides as electron-emissive material. These measures ensure a threshold energy (work function) for thermionic emission. The electrodes are the key elements of the lamp ensuring long lamp life, reliable ignition and efficient operation. Rapid warm-up is essential for fast transition from glow to arc discharge. In addition, the operating temperature should be high enough for efficient thermionic emission but low enough to prevent evaporation of the electrode material.

The tungsten electrode assembly is connected to a thin molybdenum foil, which is sealed in the silica on the opposite side. This flexible foil compensates the expansion differences of the electrode and the silica tube. The whole lamp must be designed to withstand extreme temperature gradients. Temperatures as high as 2000°C can be obtained at the tungsten rods, falling to about 500°C at the oxide coil. To avoid seal cracking, temperatures of the molybdenum foil must not exceed 250°C. Commonly, the surface of the silica jacket can be heated up to 800°C. At least 400 - 600°C jacket temperature are needed to keep the mercury vapour pressure high enough.

Around the electrodes the lamp jacket is coated with a reflective, heat resistant material. This keeps the tube ends warm, thereby preventing local mercury condensation. In a medium pressure mercury lamp, all mercury is evaporated. The lamp pressure (0.1 - 10 bar) is regulated by the amount of mercury fill. Argon is added as start gas. After excitation it forms a metastable state with an energy higher than the ionisation energy of mercury. Energy transfer from metastable argon leads to ionisation of mercury. The electrons produced provide a supply of charge carriers and induce ignition.

(ii) Spectral output

Figure 4 shows the spectral characteristics of a medium pressure mercury, iron and gallium doped arc lamp. It can be seen that for medium pressure lamps, the radiant power of the 365 nm line dominates over that of all other transitions.

Table III summarises electric characteristics of some medium pressure mercury arc lamps.

Table III Electric Characteristics of Some Medium Pressure Mercury Arc Lamps (*)

* Specifications from Heraeus Noblelight, Hanau, Germany

Total electric power W/cm	Arc length cm	Lamp power kW	Lamp voltage V	Lamp current A
120	12.5	1.5	130	12.8
120	175	21	2100	11
160	10.6	1.6	230	7.5
160	140	23	1830	14
200-250	25	5	430	13
200-250	112	22.4	1500	17

As shown in Figure 4, the addition of metal halides results in a shift of the spectral output towards longer wavelengths.

The total radiant power of a medium pressure mercury lamp is the sum of contributions from ultraviolet, visible and infrared radiations. About 15% of the total radiant power is emitted in the wavelenth range from 750 to 2000 nm. Infrared radiation contributes to the heat transfer to coating and substrate. In UV curing, temperature increase in the coated layer can induce higher conversion, but can also lead to problems when heat sensitive substrates are treated or machine components exposed.

Table IV contains the explanation of some terms used in connection with geometry, electric power, radiant power and spectral efficiency of medium pressure mercury lamps.

Table IV Terms Relating to Mercury Arc lamps

Term	Dimension	Explanation	Typical value for a medium pressure mercury arc lamp	Typical value for a microwave-powered medium pressure mercury lamp	Typical value for a low pressure mercury lamp
Arc (plasma) length	cm	Physical length of the emitting plasma	A few centimetres up to more than two metres	25 cm, several lamps can be placed in series	A few centimetres up to 1.7 m
Total electric power	$W\ cm^{-1}$	AC electric power of the lamp	40-250	120-250	0.8-10
Total radiant power	$W\ cm^{-1}$	Total power emitted by the lamp into UV, VIS and IR range	32-200	90-188	0.3-1.0
Total spectral efficiency	%	Total radiant power divided by total electric power	80	75 (45)*	7-50
UV VIS IR spectral efficiency	%	UV VIS IR radiant power divided by total electric power	25 *** 15 40	32** 25 18	5-40 1-10 -

*   Magnetron losses taken into account
**   R. Denney, Proceedings RadTech Europe`95, p.363 (1995)
***   Heraeus Instruments, Hanau Light Sources

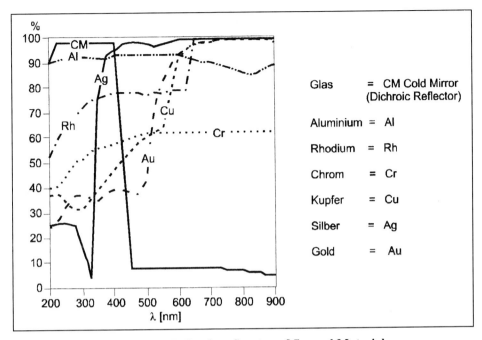

Figure 10 Reflection Spectra of Several Materials

(iii) Reflectors and lamp cooling

Reflectors are used to direct the radiated energy from the lamp towards the substrate plane. Manufactured in extruded aluminium to varying lengths and different shapes, they are also designed to provide lamp cooling. Electrode seals must be kept below 250°C, the lamp jacket needs to be kept between 600 and 800°C, and aluminium starts to deteriorate above 250°C. Air- and/or water-cooled systems are commonly used which help to cool the lamp seals and the quartz jacket.

Reflecting surfaces are usually made of highly polished and corrosion protected aluminium or stainless steel sections. To increase the efficiency of UV irradiation, reflector material with a suitable reflection spectrum is chosen. As Figure 10 shows, the reflection spectrum of aluminium or chromium gives better UV reflection than silver or gold.

Only a perfectly smooth surface shows directed (or specular) reflection where the angle of reflectance equals the angle of incidence. In practice, the non-avoidable micro-roughness of the surface leads to some spreading of the reflected light over a finite range of angles.

When it is related to UV, reflection in the visible and infrared regions of the spectrum is higher. In particular, the heat generating the infrared part is strongly reflected. This can be avoided, at least to a certain extent, when dichroic reflectors are used. A good dichroic system should drastically reduce much of the infrared and visible light energy.

A material is referred to as dichroic if the ratio of absorbed to reflected light depends on the wavelength. Dichroism is commonly observed at double refracting crystals. However, transparent dielectric multilayer systems can be constructed in which specific wavelengths are reflected and others transmitted. To prevent reflectance from lenses, such $\lambda 4^{-1}$ layers, showing cumulative constructive interference, are frequently used in optical devices.

Figure 11 A Multilayer Dielectric Coatings as IR Absorber

Figure 11b Spectral Reflectance of a Dichroic and a Standard Aluminum Reflector

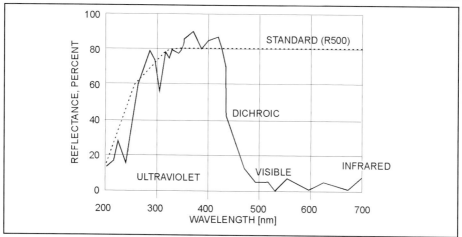

The effect of cumulative constructive interference can be also achieved by the vacuum deposition of layers, such as aluminum, titanium, cerium and silicon oxides, magnesium fluoride and zinc sulphide on transparent or opaque substrates with precisely controlled thickness. In the case of aluminum or stainless steel reflector housings, a black absorptive coating will be used as base layer (Figure 11a). Fifty or, more precisely, thickness- and refractivity-matched dielectric layers are needed to achieve the spectral reflectance characteristics given in Figure 11b [9].

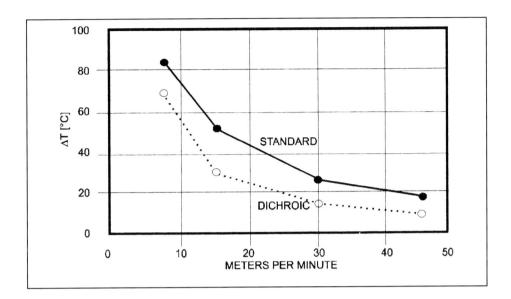

Figure 12 Effect of the Lamp Reflector on the Temperature Rise of Black Vinyl Label Stock, 240 Wcm^{-1} Lamp Power

Experimental comparison of an aluminium standard and a dichroic reflector can be made by observing the temperature rise of a substrate material passing the lamp reflector combination. Figure 12 shows the results for a black vinyl foil as substrate [9].

As shown in Figure 12, carefully constructed dichroic reflectors can provide a useful reduction in substrate temperature.

Other methods to reduce substrate heating are the filtration of infrared radiation by de-ionised and chilled water or by a suitable quartz window plate. The first method, widely applied in graphic arts, uses two quartz cooling tubes placed under the lamp between the lamp and the substrate. Besides circulating through the reflector housing, water also flows through the cooling tubes. The distilled and de-ionised water has to be kept in a closed system to avoid chemical or bacterial contamination. Water filtering can be very efficient. Up to 50% of the lamp's infrared emission can be absorbed in the filter.

Quartz filtration makes use of the effect that between 10 and 15% of infrared emission from UV lamps are absorbed by a special quartz. A quartz plate is placed between lamp and substrate and has to be air-cooled to remove the absorbed energy. In the same way as water filtration, this method reduces substrate heating but also leads to losses in UV transmission.

Using cold excimer radiation instead of filtered emission from a medium pressure mercury arc lamp is the better alternative, at least for slow-running, heat-sensitive curing applications.

In UV curing, the highest possible power per unit area incident on the substrate plane (irradiance) and a uniform irradiance across substrate travel are required to generate a high concentration of polymerisation initiating species and to minimise the effect of oxygen inhibition. A linear lamp with a reflector of a suitable and uniform cross section provides both uniformity across substrate direction and high irradiance at the substrate plane. The most simple reflector cross section normally used is an open ellipse. The ellipse has two foci. The distance between the foci is known as focal length F. Figure 13 shows an elliptical reflector as well as some formulas for the focal length F, the reflector width D at half the focal length and the minimum distance c from focal point to reflector surface.

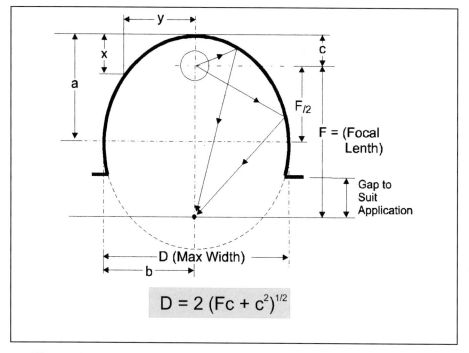

Figure 13 The Elliptic Reflector and Its Application in a Tubular Lamp

In practice, the lamp is placed on one focal point, and the substrate is passed through the other.

It is obvious that a certain amount of UV energy is re-absorbed by the lamp. Lamps with smaller diameters show less re-absorption. On the other hand, the larger the total angle around the lamp, the higher is the reflected power. For typical semi-elliptical systems a well-designed reflector can add about 30% to the irradiance produced by direct UV irradiation.

However, there are special cases such as the curing of optical fibres, where the fibre is fed through the second focus of a closed ellipse.

Another reflector cross section sometimes used is the parabola. Mathematically, it is an ellipse with the focal length of infinity. A parabola has a single focal point, and photons radiated from this point are reflected parallel to the reflector axis. Figure 14 shows a parabolic reflector as well as a formula for the reflector width D as a function of the reflector depth H and the minimum distance **A** from the focal point to the reflector surface.

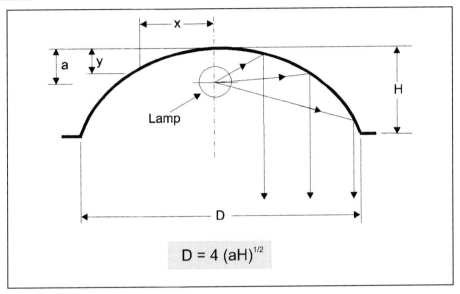

Figure 14 Parabolic Reflector

The irradiance distribution which a lamp-reflector combination produces in the substrate plane is of considerable importance. When the UV power is measured in the curing plane, elliptical reflectors produce a strongly peaked irradiance distribution while parabolic reflectors reduce and broaden the peak irradiance [10]. At equal doses (in mJ cm^{-2}), a higher irradiance (in mWcm^{-2}) is produced by elliptical reflectors.

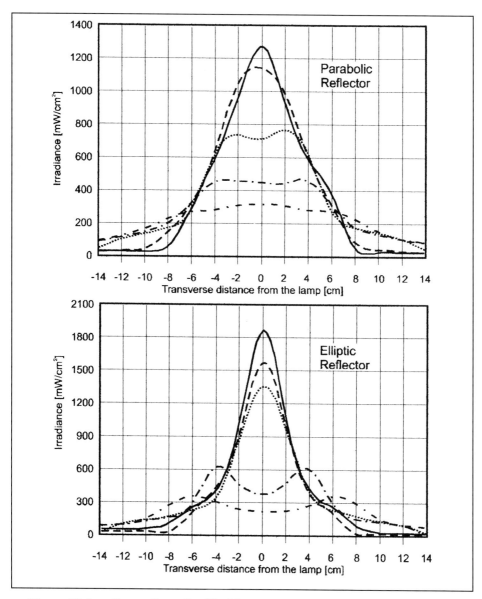

Figure 15 Irradiance Cross-Profile Measured in the Curing Plane at Different Distances form the Lamp: Elliptic and Parabolic Reflector

Practical applications prefer reflector constructions which combine simple manufacturing procedures with optimised cooling and reflecting behaviour. In addition, the reflector needs to be used as a shutter in most cases. As has been mentioned previously, medium pressure mercury arc lamps cannot be started instantaneously and need some time to be re-striked.

Interruption of lamp action would be totally unacceptable for continuous production. On the other hand, the heat from burning lamps could cause the deterioration of the substrate in line stand-by situations. Therefore, shutters are now commonly used to close over the lamp which is immediately switched to half power. Clamp operation or rotation of reflector parts are favourised shutter construction principles. An example of practically used reflector/shutter systems is shown in Figure 16. Air-or water-cooled extruded aluminum reflector housings with complicated reflecting surfaces are offered by different manufacturers.

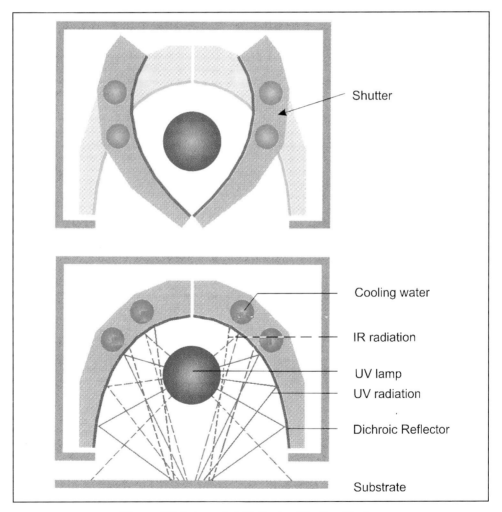

Figure 16 Integrated Reflector/Shutter Design

Although the optical design of reflectors has been thoroughly treated by Elmer [11], irradiance distribution patterns of complicated reflector cross sections should be calculated by numerical methods. Figure 17 shows an example of such a calculation.

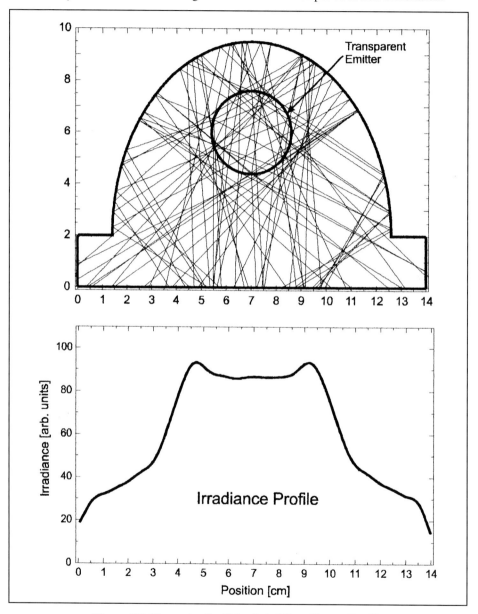

Figure 17 Numerically Calculated Irradiance Profile

(iv) Power supply and lamp control

Medium pressure mercury lamps are normally AC driven. Due to the high impedance of the lamp's plasma arc, a high electrode potential is required to maintain the arc. Potential gradients up to 30 V per cm of length are needed, and the open circuit voltage should be up to twice the operating voltage. This means that mains voltage is insufficient for practically all lamps.

- To ignite and operate the lamp, a step-up transformer is required.

There are also other important requirements with regard to the design of lamp power supply:

- Due to the negative voltage-current characteristics of the lamp discharge, a current-limiting power supply must be used.

- The burn-in characteristics of the arc lamp have to be taken into account.

Typical voltage and current time profiles as observed during the burn-in of a mercury medium pressure arc lamp is shown in Figure 18.

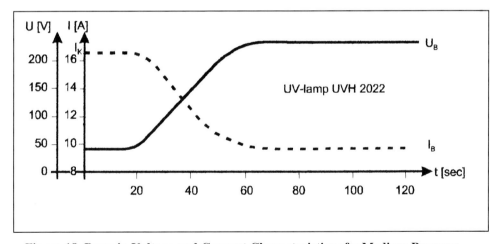

Figure 18 Burn-in Voltage and Current Characteristics of a Medium Pressure Mercury Arc Lamp

Immediately after ignition, mercury is still liquid and the discharge takes place in argon. The lamp voltage is low and the current is practically limited by the short-circuit current delivered from the power supply. With increasing temperature mercury vaporises. The lamp impedance increases. This causes the lamp voltage to increase and the current to decrease. After about one minute the burn-in period is finished and the lamp has reached stationary conditions.

A simple solution of a power supply providing sufficiently high AC voltage for ignition and lamp operation, current-limitation for the arc discharge and a limited short-circuit current at burn-in, is a step-up transformer with a lamp and "ballast" capacitor placed in the secondary circuit of the transformer (see Figure 19).

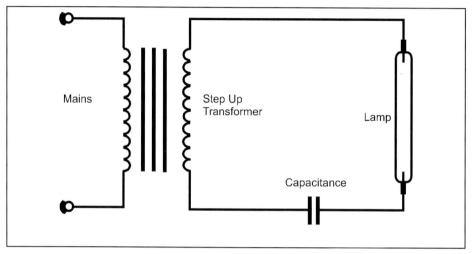

Figure 19 Power Supply for a Medium Pressure Mercury Arc Lamp: Capacitive Ballast

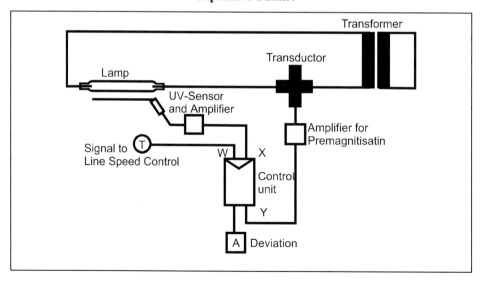

Figure 20 Transductor as Inductive Ballast and Radiant Power Control

This is the most popular power supply for running mercury arc lamps with large arc lengths. As lamp power is proportional to the capacitance, step-wise power variation (typically in the range of 40 to 100%, as long as the lamp is operated at appropriate cooling conditions) can be simply achieved by switching capacitors.

As shown in Figure 20, there is another solution which is sometimes applied for lamp power supplies: a transductor can serve as inductive ballast and current control at the same time. The current flowing through the lamp is controlled by transductor premagnitisation. This allows continuous lamp power control within the 40 and 100% level. Besides continuous current control, there are some other benefits which result in smooth lamp burn-in, a short dark period between the AC half waves, early re-ignition after each voltage half cycle, low noise generation and reliable long-term operation [12].

Lamp voltage and current waveforms, as generated by a transductor controlled power supply are schematically shown in Figure 21. The light output of the lamp Φ closely follows the lamp current I. Within each half cycle, light is emitted as long as enough current is flowing. There is no emission within a distinct phase range close to the AC voltage crossover. This "dark period" can be kept constant if transductor current control is used.

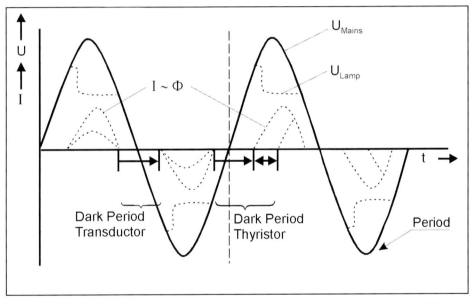

Figure 21 Lamp Voltage, Current and Emission Waveforms as Measured for a Transductor (Left) and Thyristor Control (Right) (Schematically)

Thyristors can also be applied to provide stepless control of the lamp current. In comparison to transductor control, the dark period is not constant but increases at lower power.

The power supplies as shown in Figure 20 are not stabilised, i.e. the lamp power depends on mains voltage fluctuations. Within the permitted fluctuation range of the mains voltage, the deviation from the rated power can amount up to 20%. The resulting fluctuations in irradiance may cause quality reduction in curing processes. This can be avoided by using a closed loop control circuit as shown, for example, in Figure 22 [13,14].

A UV sensor is used to monitor the radiant power of the lamp. In a control unit the sensor signal is compared with a preset signal or a signal provided by a tachometer. The resulting difference in signal is magnified and used to control the lamp current via transductor or thyristor switches.

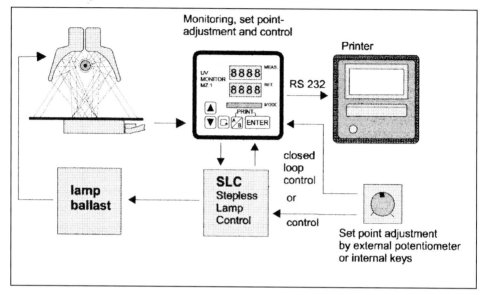

Figure 22 Closed Loop Control of a Lamp Power Supply

Recently, electronically-controlled lamp power supplies have been developed [13,14]. Compared to a conventional lamp power supply, the following benefits of an electronic power supply are mentioned [13]:

- current or power stabilised,
- electrical efficiency >92%,
- power variation possible from 12 to 108%,
- very low stand-by power adjustable,
- compact dimensions and less weight.

The electronic power supply is operated at medium frequency and is, at present, limited to a maximum power of about 5 kW. A block diagram illustrating the main components of such a power supply is given in Figure 23a. Figure 23b shows the UV-C irradiance of a lamp driven by the electronic power supply as a function of electric power.

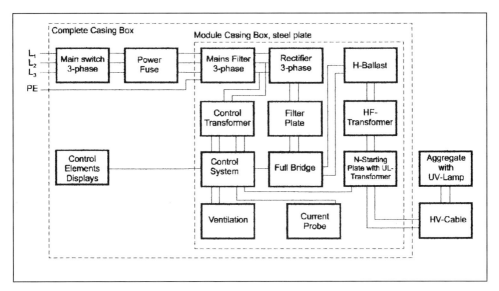

Figure 23a Electronically-Controlled Lamp Power Supply

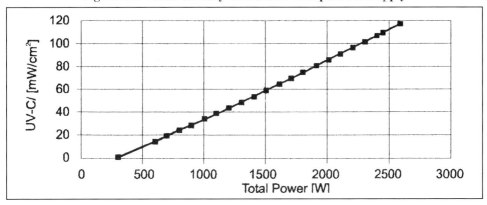

Figure 23b UV-C Irradiance of a Lamp Controlled by an Electronic Power Supply

(v) **Ozone-free lamps**

Oxygen in the atmosphere absorbs UV radiation below about 225 nm to produce ozone, which is corrosive and toxic. As shown in Figure 4, medium pressure mercury lamps show a spectral continuum below 240 nm, part of which contributes to ozone generation. The formation of ozone can be prevented by the use of special silica as lamp jacket, which is doped to cut off ozone generating wavelengths. Because the cut off filter profile of the doped silica is not a step function, but rises smoothly from about 240 to 300 nm, the UV output is reduced in this wavelength range. This causes considerable reduction in curing efficiency, which in most cases is not tolerated by the user. Ozone removal by exhausting and filtration is a more common solution.

(vi) Lamp life

The deterioration processes which take place in a medium pressure mercury arc lamp are mainly due to material evaporation or sputtering from the electrodes. Tungsten or other materials are then deposited inside the lamp jacket and reduce its transmission. Because sputtering of the material takes place mainly during the burn-in period of a lamp, the number of lamp on-off cycles affects lamp life in a serious manner. The quality of the silica used for the lamp jacket is also of great importance for lamp life. A low content of hydroxyl groups hinders water permeation into the lamp, and a low level of metal impurities prevents solarisation. There are some other measures which can be taken to prolong lamp life: higher argon pressure to reduce sputtering and the addition of halogens to remove tungsten from the tube wall, which is then redeposited on the electrodes (halogen cycle).

Lamp aging will cause certain changes in the spectrum. In comparison with the visible part of the spectrum, the UV radiant power decreases more rapidly. This should be taken into account when the curing efficiency of a lamp is judged as a function of time.

As an example, the power loss as a function of time measured for a 25 cm wide 300 Wcm^{-1} medium pressure arc lamp is illustrated in Figure 24 [15].

The percentage of power loss over time is clearly higher for 254 than for the other wavelengths monitored. The lowest degradation was observed for the 365 nm line.

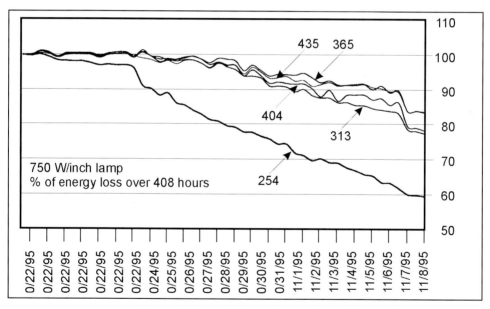

Figure 24 Loss of Lamp Power Over Time

Manufacturers often quote a "guaranteed" lamp life which better should be defined as "the time the lamp can be used for a certain curing application". For general curing applications, where a 25% fading of the lamp radiant power can be tolerated, lamps can be used for 1000 to 1500 hours.

3. Microwave-powered medium pressure mercury lamps: Fusion UV Systems F300, F450 and F600

(i) Construction of the irradiator

Since the end of the 1940s, small electrodeless microwave-excited lamps have been used for special applications in optical spectroscopy, analytical chemistry, UV and VUV photochemistry [16].

However, the lamps were designed almost exclusively to produce low-powered monochromatic emission.

As mentioned above, the extension of the basic concept of microwave powering to medium pressure mercury lamps was pioneered by researchers and engineers of Fusion UV Systems Inc. Gaithersburg, Maryland USA.

A description of the principles of operation has been given in the US Patents 4, 043, 850, 3, 911, 318, 3, 872, 349 and others. The irradiator contains two magnetrons as microwave generators, two waveguides and a closed non-resonant microwave chamber formed by a semi-elliptical metallic reflector with flat ends and a fine metal mesh. The lamp which is a closed silica tube, is placed with its axis in the focus of the reflector ellipse. It forms a highly dissipative load for the microwave circuit (see Figure 25). The first commercialised irradiator was a 300W/inch version. Since 1990, a more powerful 600W/inch irradiator has been on the market and described in a number of papers [17,18,19].

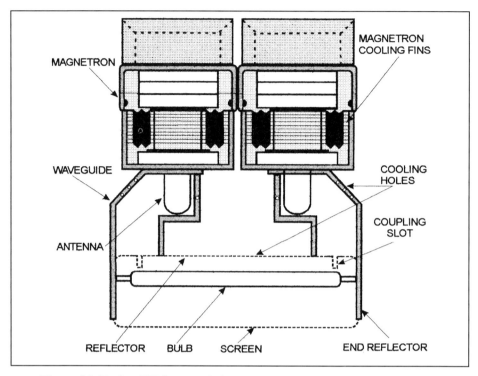

Figure 25 Fusion UV Systems Microwave-Powered Electrodeless Medium Pressure Mercury Lamp

Microwave energy is generated by two 1.5 or 3.0 kW 2450 MHz magnetrons and fed through an antenna system to each waveguide. The metalic reflector and the metallic screen closure mesh form a microwave chamber which, from the electrical point of view, has a non-cavity structure. The microwave energy coupled into the chamber by precisely designed slots is uniformly distributed over the lamp tube. Standing waves which would result in a resonant distribution of microwave energy are avoided by slot design and choosing pairs of magnetrons slightly differing in frequency. Magnetrons and waveguides are cooled by filtered air which is also passed into the microwave chamber through small holes in the reflector. This downward air flow serves to cool the lamp bulb. In addition, it helps to keep lamp and reflector clean.

The lamp bulb is made of vitreous silica and has an inner diameter of 9 mm for 300 $Winch^{-1}$ or 13 mm for 600 $Winch^{-1}$ irradiators. The bulb length is 9.4 inch (23.9 cm) for the F 450 (see Figure 26) and F 600 or 6 inch (15.1 cm) for the HP-6 system.

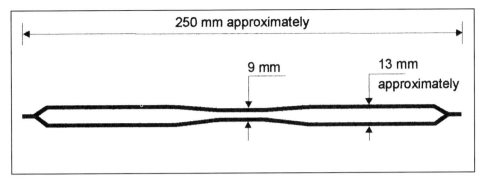

Figure 26 Electrodeless Lamp

The electrodeless lamp is generally filled with mercury and a neutral starter gas. It runs at mercury pressures of 5 to 20 bar which is higher than for arc lamps (1-2 bar). Special materials such as metal halides can be added to the gas fill in controlled quantities. Energy transfer to the metals by high energy photons emitted from mercury and by collisions with excited mercury atoms lead to excitation of the additive resulting in a characteristic "fluorescence" emission. The frequently used D or V lamps contain iron and gallium, respectively.

Some basic characteristics of Fusion UV Systems microwave-powered UV irradiators are summarised in Table V.

Table V Characteristics of Microwave-Powered UV-irradiators

Peak Irradiance as a Function of Bulb Diameter and Input Power			
Lamp System	Bulb Type	BulbDiameter mm	Peak Irradiance $Wcm^{-2}(UVA_{EIT})$
F-450 (350 W/inch) Control	H D V	9 9 9	2.01 5.23 1.84
F-450 (350 W/inch)	D D D	13 11 9	4.39 4.88 5.23
F-600 (560 W/inch)	H D V	13 13 13	2.66 7.46 2.38
F-600 (560 W/inch)	D D D	13 11 9	7.46 9.67 9.88
Electrode Lamp (reference) 300 W/inch 400 W/inch	Hg Hg	25 25	1.35 2.15
Note: Radiometer readings taken at 20 fpm at lamp center using EIT UV Powerpack, 10 watt/cm² range, UVA band (365±30 nm)			

The irradiance pattern observed in a certain intercepting irradiation plane is affected by the reflector cross section, the bulb dimension and the distance of the plane from the irradiator face.

The semi-elliptical reflector used for all electrodeless Fusion lamps has its focus 2.1-inch (5.33 cm) from the irradiator face. A typical irradiance distribution as obtained in the focal plane is shown in Figure 27. It compares a lamp with a 13 mm bulb operating at 600 and 400 $Winch^{-1}$ with a 9 mm bulb lamp operating at 300 $Winch^{-1}$. The highest peak irradiance is obtained in the focal plane. As expected, the light from the smaller diameter bulb is focused more finely.

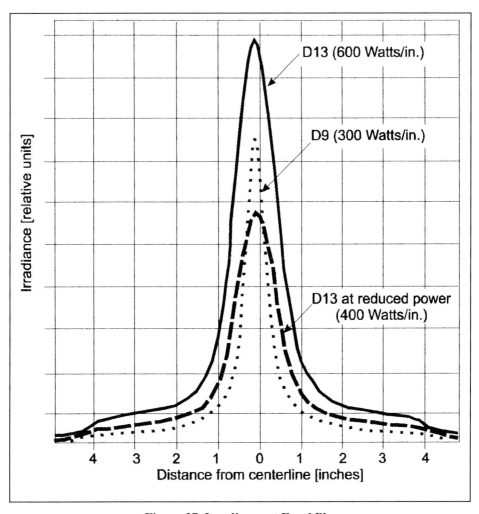

Figure 27 Irradiance at Focal Plane

(ii) Spectral output

Figure 5 shows the spectral output of microwave-powered irradiators using H, D and V bulbs.

The spectral output is given as radiant power (in W/inch) in wavelength intervals of 10 nm. By summarising the radiant power of the 10 nm intervals belonging to a certain central wavelength, a spectral output chart can be obtained which is easy to analyse. Typical UV and IR radiant powers as measured for different H and D bulbs are summarised in Table VI.

Table VI Radiant Power of Different H and D Bulbs*

* Specifications from Fusion UV Systems, Gaithersburg, USA

Bulb type and power	Radiant power (Wcm^{-1}) UV (200-450 nm)	Radiant Power (Wcm^{-1}) IR (700-2500 nm)
H 120 Wcm^{-1}	38	22
D 120 Wcm^{-1}	75	22
H 240 Wcm^{-1}	58	45
D 240 Wcm^{-1}	120	45
Medium pressure Hg arc lamp 120 Wcm^{-1}	48	50

The spectral output distributions of a medium pressure mercury arc lamp and a corresponding microwave-powered lamp have been compared by Philips [3]. His conclusion was that an arc and an electrodeless lamp of the same power input between 200 and 600 nm produce about the same total radiant power.

(iii) Power supply

The power supply contains the high voltage generators for the magnetrons, filament heating transformers, a control board with status and fault indicators and numerous safety and production interlocks. For the F450 and F600 series a variable power supply (25 to 100%) is available. A microwave detector placed close to the irradiator will automatically switch off the system if the microwave leakage exceeds a given threshold.

After starting the system, it takes about 10 s until the magnetron filaments are warmed up. High voltage is then supplied to the magnetrons and microwave energy is fed into the waveguide. A small part of the microwave energy is absorbed by a probe which is used to ignite a small mercury low pressure lamp located behind a reflector hole. Absorption of 254 nm photons by mercury droplets causes photoelectron emission. The electrons emitted are accelerated by the microwave field, rapid charge carrier generation takes place and mercury evaporates. About 3 s after ignition, full light output is obtained.

Multiple power supply configurations are available for irradiator arrays.

(iv) Electrodeless vs. arc lamp: comparison of technical characteristics

Table VII attempts to compare technical characteristics of Fusion UV Systems electrodeless lamps with those of medium pressure mercury arc lamps.

Table VII Electrodeless vs. arc lamp: comparison of technical characteristics

Parameter	Electrodeless lamp	Arc lamp
Maximum electric power (Wcm^{-1})	240	240
Power regulation and control	comparable	
Electric efficiency (%)	50-60	80-90
Total UV radiant power	comparable	
IR output power	less	more
Peak irradiance	high	moderate
Irradiation length (cm)	25, for multiple irradiator units length comparable to* arc lamps	up to 250
Metal halide lamp fills available	comparable	
Start and restrike time	short	long
Shutter needed to protect the substrate	no	yes
Power regulation and control	comparable	
Lamp lifetime	more	less
Cooling	air	air or water
Irradiator dimensions	larger	smaller

(v) Fusion UV Systems Versatile Irradiance Platform (VIP)

In 1997, Fusion UV Systems introduced the latest 10 inch microwave-powered lamp system with variable output power and interchangeable electrodeless bulbs. The modular lamp system consists of an irradiator, a variable power supply, an I/O module, a microwave detector and a blower for air cooling. Electronic ignition of the bulbs (see Figure 13 of Chaper IV) allows higher bulb pressures. The maximum specific electrical power is 240 Wcm^{-1}. It can be controlled from 25 to 100% in 1% increments. It is Fusion's intention to develop high-powered bulbs with emissions tailored closely to match the absorption bands of common photointiators. At present, three interchangeable bulbs are available: VIP 308, VIP 535 and VIP Cobalt, some others are in preparation.

The VIP 308 bulb is a high-pressure xenon chloride excimer lamp with most of its energy concentrated in the wavelength range from 280 to 310 nm. Its emission spectrum is shown in Figure 14, Chapter IV.

The VIP 535 bulb is an excimer lamp using copper bromide as emitter. Its maximum radiant power is in the wavelenth range between 530 and 540 nm, but there are also some weaker lines between 250 and 440 nm (see Figure 28, Upper Curve). The VIP Cobalt bulb has a very high radiant power between 340 and 360 nm (see Figure 28, Lower

Curve). Its total UV/VIS radiant power is comparable to the Fusion D bulb, but its emission spectrum better matches the absorption band of such reactive photinitiators as Irgacure 369 and 651.

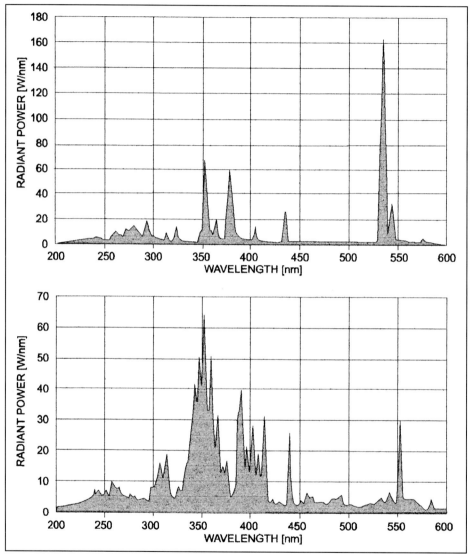

Figure 28 Radiant Power as Function of Wavelength: VIP 535 (Upper Curve), VIP Cobalt (Lower Curve)

McCartney and Bao[20] compared the radiant power emitted into a 30 nm wavelength band of F600 H, D and V bulbs with that of new, currently still experimental, H, D, and V bulbs made for the VIP irradiator. The new bulbs could be run with enhanced radiant powers of 30 to 40%. Figure 29 shows a comparison of the bulb radiant powers and also notes an experimental krypton chloride (222 nm) excimer bulb.

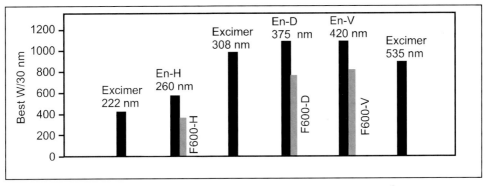

Figure 29 Radiant Power from F600 and VIP (En) 600 W inch^{-1} bulbs

4. Mercury and xenon short arc lamps

As the name implies, short arc lamps are designed to work as continuous quasi point light sources. Xenon lamps are operated at xenon pressures of about 20 bar. A high-current driven plasma discharge is compressed into an arc length of a few millimetres. Very intense UV, visible and infrared emissions are produced. Figure 30 illustrates construction and relative radiant power distribution of a typical short arc xenon lamp.

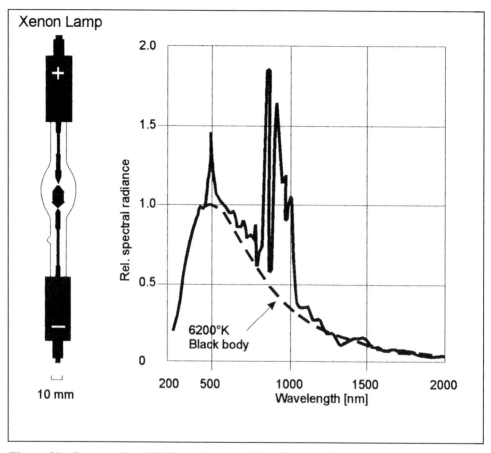

Figure 30 Construction (Left) and Spectral Distribution (Right) of a Xenon Short Arc Lamp

The spectral output shows a nearly continuous distribution with smaller lines superimposed in the visible and more pronounced broad lines in the near infrared. In the visible wavelength range, this spectral distribution closely resembles that of a black body radiator of 6200 K, i.e. the colour quality of a xenon short arc lamp is close to sun-light. The construction of mercury short arc lamps is very similar to that of the corresponding xenon lamps. Their emission spectrum is comparable to that of medium pressure mercury arc lamps (see Figure 4). In comparison to xenon lamps, the relative UV radiant power is higher and special "deep UV" sources are available.

In curing applications, xenon and mercury short arc lamps are used as high-intensity light sources for spot curing. Such systems enable the user to deliver an intense spot of energy to a targeted area and to accomplish curing of complex three dimensional objects. Flexible light guides are used to transmit the light from the arc of the lamp to precisely where it is needed [21].

Figure 31 shows the optical components and the light path of a typical spot curing system. Spot curing units are compact and fit easily into manufacturing processes. Illumination conditions can be preset using filters and shutter control. Main application fields are spot curing of adhesives and stereolithography.

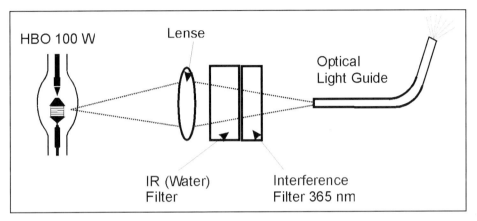

Figure 31 Spot Curing Device

IV. REFERENCES

1. H. Yasuda: Plasma Polymerization, Academic Press, Orlando, London (1985)
2. Venugopalan (ed.): Reactions under Plasma Conditions, Wiley, New York (1971)
3. R. Phillips: Sources and Applications of Ultraviolet Radiation, Academic Press, London, New York (1983)
4. P.K.T. Oldring (ed.): Chemistry & Technolgy of UV&EB Formulation for Coatings, Inks and Paints, Vol.1, SITA Technology, London (1991)
5. American Institute of Physics Handbook, D.E.Gray(ed.), Mc-Graw Hill, New York (1982)
6. US Patents No. 3,872,349; 3,911,318; 4,042,850 and others
7. B.J. Eastlund, M.Ury, C.H.Wood, SME Papers Ser. 75-301 (1975)
8. L.S. Levine, Radiat.Phys. Chem. 9, 819 (1977)
9. R.W. Stowe, Proceedings RadTech North America `92, p.447 (1992)
10. B. Schaeffer, S. Jönsson, M.R. Amin, Proceedings RadTech Noth America `94, p. 314 (1994)
11. W.B. Elmer: "The Optical Design of Reflectors", 2nd. edn., Wiley-Interscience, New York (1980)
12. P. Klamann, private communication
13. P. Beying, Proceedings RadTech Europe`97, p. 77 (1997)
14. DPL data sheet 1997
15. S.B. Siegel, P.Mandellos, D.Luster, Proceedings RadTech North America `96, p. 263 (1996)
16. A.A. Lamola (ed.): Creation and Detection of the Excited State, Vol.I, Part A, Marcel Dekker (1971)
17. R.W. Stowe, Proceedings RadTech `90 North America, Vol.1, p. 165 (1990)
18. R.W. Stowe, Proceedings RadTech `90 North America, Vol.2, p. 173 (1990)
19. W.R. Schaeffer, Proceedings RadTech `92 North America, Vol.1, p. 201 (1992)
20. R. McCartney, R.Bao, Proceedings RadTech Europe`97, p.83 (1997)
21. R. Hood, Proceedings RadTech `90 North America, p. 159 (1990)

CHAPTER IV

UV CURING EQUIPMENT – MONOCHROMATIC UV LAMPS

I. MONOCHROMATIC UV RADIATION FOR CURING

Until the mid-90's, no monochromatic UV source was available which could compete in cure speed and curing performance with the polychromatic medium pressure mercury arc lamps, which were able to achieve a high UV radiant power of up to 40 W/cm with an excellent technical standard for the lamp and related curing equipment. Moreover, photoinitiator absorption was matched to the mercury emission lines in a nearly perfect manner. As a result, radical generation rates were obtained high enough rapidly to overcome oxygen inhibition in the layer to be cured and to enable fast cure speeds. Even in cationic systems a proton concentration was generated which allowed a reasonable cure speed. Table 1 shows that commercial low pressure mercury lamps have an unacceptable low UV output for general curing applications. Lamps using rare gas resonance emission are even less powerful. Additionally, in liquid acrylates or epoxides the penetration depth of their high energy photons is limited to less than one micron. Excimer and nitrogen laser are pulsed UV sources with a variety of interesting wavelengths and high output peak powers. However, they cannot compete with mercury arc lamps in large area, fast cure applications due to their low pulse repetition rates.

Interesting new possibilities were opened up with the development of a completely new class of UV sources - the excimer lamp [1].

The term excimer radiation is normally used in connection with excimer lasers. Excimers (**exc**ited di**mer**s, tri**mer**s) are weakly bound excited states of molecules that do not posses a stable molecular ground state [2].

The most important commercial excimers are formed by electronic excitation of rare gases (He_2^*, Ne_2^*, Ar_2^*, Kr_2^* Xe_2^*), rare gas-halides (ArF^*, KrF^*, $XeCl^*$, XeF^* etc), halogens (F_2, Cl_2, Br_2, I_2) and mercury halogen mixtures (HgCl, HgBr, HgI). They are unstable and decay by spontaneous optical emission.

Stimulated (laser) excimer emission can be generated in pulsed high pressure glow discharges. On the other hand, dielectric barrier (silent) discharges [1] or microwave discharges [3] can be used to produce quasistationary or continuous incoherent excimer radiation.

II. UV AND VISIBLE EMISSION FROM EXCIMERS

1. Emission from dielectric barrier discharges in rare gases and rare gas halide mixtures

(i) Dielectric barrier discharge

For efficient dielectric barrier discharge excimer formation, electrons (e$^-$) of at least 10 to 25 eV energy are required to create excited rare gas atoms (Rg*) or ions (Rg$^+$).

Addition of low energy electrons to halogen molecules and dissociation of the transient molecular halogen anion leads to the formation of halogen anions (Hal$^-$).

$$e^- + Rg \rightarrow Rg^*, Rg^+, \qquad e^- + Hal_2 \rightarrow Hal^- + Hal \qquad (1)$$

Table I Monochromatic UV Sources and their Main Characteristics

Commercial monochromatic UV sources	Emission generated by	Main emission wavelengths	UV Radiant power	Main applications	Remarks
Low pressure mercury lamp	low pressure glow discharge	254 and 185 nm	0.1-1 Wcm^{-1}	water disinfection, preparative photochemistry	weaker emission lines from 313 to 578 nm
Excimer laser Nitrogen laser	pulsed discharges	158, 193, 248, 308, 350 nm 337 nm	up to 10^8 W peak pulse power, 10^2 W average power	photophysical and photochemical research	short pulse (5-50 ns) operation only
Excimer lamps	dielectric barriere (silent) discharge microwave discharge low pressure high pressure	172, 222, 308 nm 222, 308 nm	1-10 Wcm^{-1} 1-5 Wcm^{-1} 12-64Wcm^{-1}	UV curing, material deposition, surface modification, dry etching, pollution control, plasma display, preparative photochemistry	strong R&D activities to increase the UV radiant power and to provide other excimer transitions for technical use

If the gas pressure is high enough, rare gas excimers Rg$_2$* are formed in a subsequent three body reaction:

$$Rg^* + Rg + Rg \rightarrow Rg_2^* + Rg \qquad (2)$$

Rare gas/halogen exciplexes (for simplicity also called excimers) RgHal* are predominantly generated by ion recombination:

$$Rg^+ + Hal^- + Rg \rightarrow RgHal^* + Rg \qquad (3)$$

In practice, electron energies larger than 10 eV are required to produce the precursors of excimers, while pressures in the range of 1 bar are needed to prevent their deactivation before excimer formation can occur.

Dielectric barrier discharges, typically operated at high voltages (5-10 kV) and frequencies below 1 MHz, are ideally suited to generate, at normal pressure, the electron energy distribution needed

A dielectric barrier discharge configuration can be characterized as a cylindrical, planar or parallel plate capacitor where one or both of the electrodes are covered by a dielectric. A typical cylindrical discharge lamp suitable for excimer UV generation [4] is shown in Figure 1.

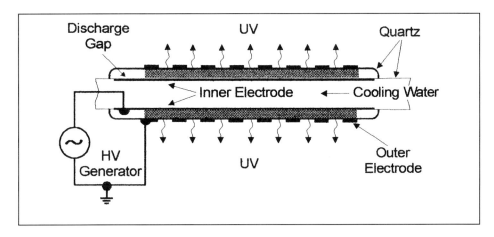

Figure 1 Cylindrical Excimer Lamp Configuration with Annular Discharge Gap and External Electrodes

When the high voltage applied to the discharge gap exceeds the breakthrough voltage, a current starts to flow, which, after a short time is switched off by the decrease of the electrical field strength in the gap caused by charging the dielectric surface. The total duration of the current pulse is a few nanoseconds and excimer emission may occur in this time. A schematic representation of the voltage current characteristics and the pulsed excimer emission from a barrier discharge is given in Figure 2.

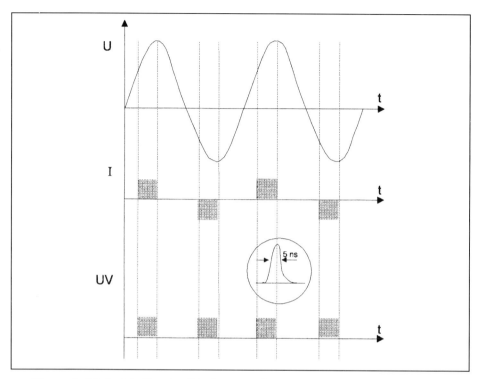

Figure 2 Dielectric Barrier Discharge - Time Profile of the High Voltage (U), the Current Pulse (I) and the Excimer Emission (UV)

A very important feature of the barrier discharge is that the current flow through the gap is generated by a large number of randomly distributed microdischarges. The microdischarge filaments can be characterized as weakly ionised plasma channels with a filament radius of about 0.1 mm, a current density of 10^2 to 10^3 Acm^{-2}, an energy density of 1-10 mJcm^{-3} and a dissipated total energy of about 1 μJ [5].

In this respect the microdischarge filaments show characteristics resembling those of a transient high pressure glow discharge. The microdischarge conditions are comparable to those of discharge pumped excimer lasers. Thus, the excimer radiation of a discharge lamp consists of a large number of pulsed "microlasers". This makes them very suitable as high power UV emission sources.

As in the case of lasers, the physico-chemical mechanism leading to the population of the emitting excimer state is called "pumping" mechanism.

(ii)   **Rare gas excimers**

As an example, the pumping mechanism involved in the formation of xenon excimers is schematicallly illustrated in Figure 3 [1,2,6].

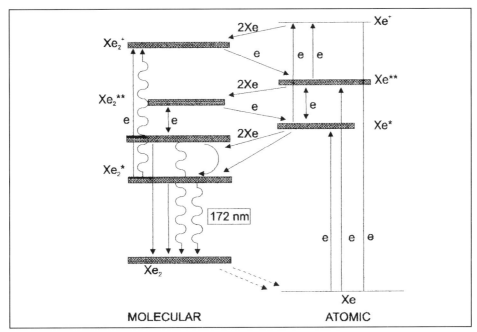

Figure 3 Simplified Pumping Scheme for Xenon

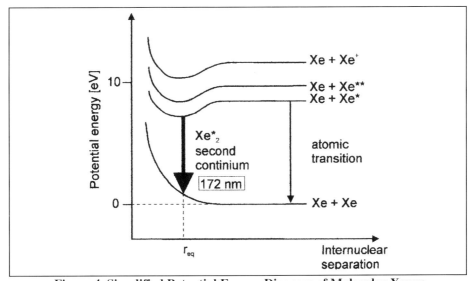

Figure 4 Simplified Potential Energy Diagram of Molecular Xenon

The simple overall reaction scheme (1) to (3) also applies to xenon. For reasons of simplicity, the various excited states are summarised as two; excited atomic (Xe*,Xe**) and molecular (Xe_2*, Xe_2**) states. Electrons generated during microdischarge excite or ionise the xenon atoms:

$$e^- + Xe \rightarrow Xe^*, Xe^+ \quad (4)$$

At pressures above 50 mbar the formation of molecular ions proceeds very rapidly and via recombination contributes to the formation of higher excited atomic states Xe**:

$$Xe^+ + 2Xe \rightarrow Xe_2^+ + Xe \quad (5)$$
$$Xe_2^+ + e^- \rightarrow Xe^{**} + Xe \quad (6)$$

The molecular excited state Xe_2* is essentially formed in a three body reaction of an electronically excited xenon atom with two others in the ground state:

$$Xe^* + 2Xe \rightarrow Xe_2^* + Xe \quad (7)$$

The Xe_2^* excimer emits a 172 nm photon. The resulting molecular ground state is unstable and dissociates into two xenon atoms:

$$Xe_2^* \rightarrow 2\ Xe + h\nu\ (172\ nm) \quad (8).$$

The radiation process (8) is in competition with the quenching of Xe_2^* through collisions with xenon atoms and by other Xe_2^* molecules. This leads to a decrease of the emission efficiency at higher pressures.

The potential energy diagram of molecular xenon as given in Figure 4 illustrates the weakly bound excited and ionic xenon molecular states, the emitting excimer state and the decaying molecular ground state. The atomic transition indicated is the xenon resonance line.

Figure 5 shows how a dielectric discharge in xenon leads from the generation of resonance to excimer radiation at increasing pressure [7]. Similar spectra were obtained for Ar_2* (126 nm) and Kr_2* (146 nm).

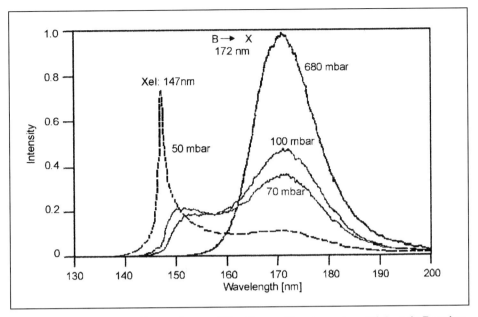

Figure 5 Pressure Dependence of the Xenon Spectrum in a Dielectric Barrier Discharge

(iii) Rare gas halide exciplexes (excimers)

Rare gas halide excimer transitions are widely used in commercial lasers [2,8] and in some cases for excimer lamps.

A large number of excimer transitions is available in the UV and visible range of the spectrum. They form an important potential for the further development of monochromatic excimer lamps whose emission could cover the UV-VIS range by narrow steps.

Similar to the case of rare gases, the potential energy diagram of rare gas halides shows decaying ground states (X, A) and bound excited states (C, B, D).

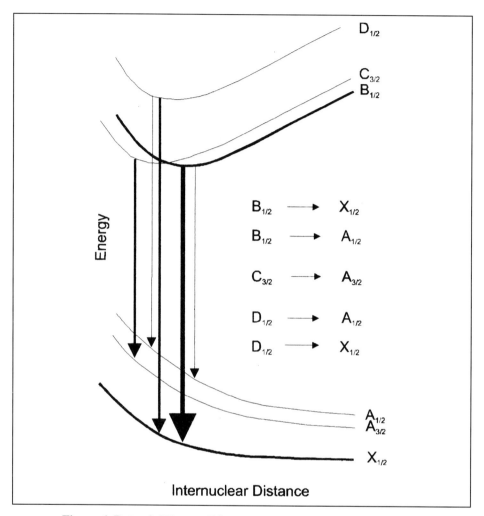

Figure 6 Potential Energy Diagram of Rare Gas Halogen Molecules

The splitting of the molecular ground state into three levels (labelled as $X_{1/2}$, $A_{3/2}$ and $A_{1/2}$) is due to the combination of the rare gas ground state ^1S and the spin-orbit splitting of the ^2P levels of the halogen atom. The first excited bound states (labelled as $C_{3/2}$, $B_{1/2}$ and $D_{1/2}$) are generated by a positive rare gas ion ^2P Rg^+ and a negative halogen ion ^1S X^- (X=F,Cl,Br,I). Possible transitions are indicated in Figure 6. Intensity as well as spectral half-width of the radiation generated from various transitions differ very much. Typical strong (laser) transitions are B-X transitions. D-X and D-A transitions are much weaker. B-A and C-A transitions are weak and broad. An emission spectrum observed from a microwave excited XeCl excimer lamp (Figure 7) illustrates this behaviour [9].

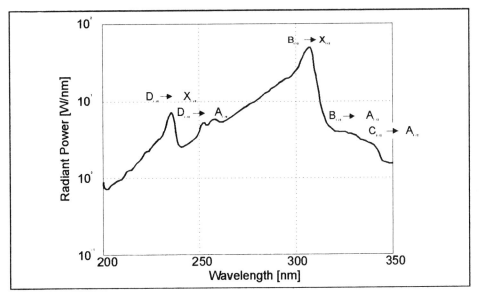

Figure 7 Excimer Emission Bands Observed from a Microwave Excited XeCl Lamp

Table II summarises the main peak wavelengths of rare gas halide excimers (B-X transitions).

Table II Main Peak Wavelengths of Rare Gas Halide Excimers [2]

Excimer	Main peak wavelength (nm)
NeF	108
ArF	193
ArCl	175
ArBr	165
KrF	248
KrCl	222*
KrBr	207
KrI	190
XeF	351
XeCl	308*
XeBr	282(*)
XeI	253

* excimer lamp commercialised

A large number of atomic, molecular and ionic species are involved in rare gas halide excimer formation. Energetic electrons formed in the microdischarge ionise and excite rare gas and generate halogen species in a dissociative electron capture reaction:

$$e^- + Rg \rightarrow Rg^*, Rg^+, \quad e^- + X_2 \rightarrow X^- + X, \quad (9)$$
$$(Rg = Ne, Ar, Kr, Xe) \quad (X = F, Cl, Br, I)$$

The main pathway of RgX* exciplex formation is a three body recombination of positive rare gas ions with the negative halide ions:

$$Rg^+ + X^- + M \rightarrow RgX^* + M \quad (10)$$

M is a collisional third partner e.g. from the buffer gas. Additionally, exciplexes can be formed by:

$$Rg^* + X_2 \rightarrow RgX^* + X \quad (11).$$

As in the case of rare gas excimers, the exciplex molecules formed are unstable and decompose by photon emission within a few nanoseconds:

$$RgX^* \rightarrow Rg + X + h\nu \quad (12).$$

The radiative process competes with several quenching processes. The most important quenching process is a three body reaction forming triatomic species at high pressures:

$$RgX^* + 2Rg \rightarrow Rg_2X^* + Rg \quad (13).$$

(iv) Halogen Excimers

Halogen excimers were studied to understand the pumping mechanism in the corresponding lasers. However, excimer radiation from F_2, Cl_2, Br_2 and I_2 was also generated using dielectric barrier discharges in halogen/helium or argon mixtures [7]. Their peak wavelengths are 158 nm for F_2, 259 nm for Cl_2, 289 nm for Br_2 and 342 nm for I_2.

(v) Mercury vapour excimers

There exists a strong similarity of mercury halogen molecules with the rare gas halogen system. Mercury atoms behave similar to rare gas atoms. The typical weakly bound ground states and the stronger bound excited states of ionic character are found there as well. Thus, nearly the same type of excimer transitions is observed: HgI: 443 nm, HgBr: 503 nm and HgCl: 558 nm.

(vi) General characteristics of incoherent excimer radiation

Excimers as radiation sources exhibit very special characteristics:

- Excimers can be very efficient energy converters. Theoretical predictions for Xe_2^* or Kr_2^* excimers said that 40-50% of the energy deposited by the electrons is converted to radiation [1, 10]
- Due to the absence of a stable ground state, no self absorption of the excimer radiation can occur.
- In most cases low pressure excimer UV sources show a dominant narrow band (1-3 nm) transition. They can be regarded as quasimonochromatic.
- Excimer systems can be pumped with very high power densities. Thus extremely bright UV sources can be built.
- For industrial applications of excimer UV sources, the dielectric barrier and the microwave discharge are simple, reliable and efficient excitation modes.
- The large number of VUV, UV and VIS excimer transitions available (see Figure 8) allow selective photoexcitation for many systems.

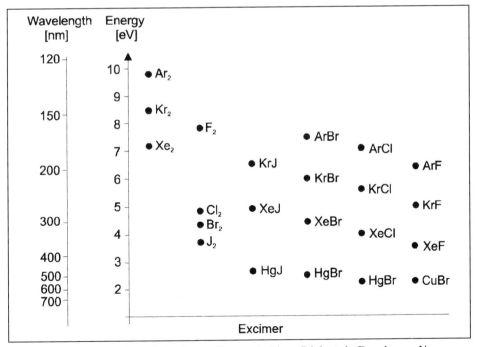

Figure 8 Selection of Excimers Generated in a Dielectric Barrier and/or a Microwave Discharge

2. Commercial barrier discharge driven excimer lamps

In 1988 dielectric barrier discharge driven excimer lamps were first described by Eliasson and Kogelschatz [1] from Asea Brown Boveri (ABB) Baden (Switzerland). The authors developed a cylindrical excimer lamp configuration with an annular discharge gap and external electrodes (see Figure 1) and demonstrated the efficient and reliable operation of the lamp as excimer UV source for xenon (172 nm), krypton chloride (222 nm) and xenon chloride (308 nm). First successful efforts were made to use excimer lamps for applications such as photodegradation of pollutants [11,12], metal deposition on polymers [13], surface modification of polymers [14] and the curing of printing inks [15]. In all these applications the great potential of excimer lamps as new industrial UV sources was demonstrated.

In 1993 Heraeus Noblelight (Hanau, Germany) acquired the excimer lamp technology from ABB and commercialised it.

The main lamp geometry remained cylindric with efficient water cooling inside (see Figure 9). Figure 10 shows the emission spectra of the excimer lamps. Special high purity quartzglas was used for the lamps with electrodes outside the discharge chamber. Thus, no quartz to metal seals were needed.

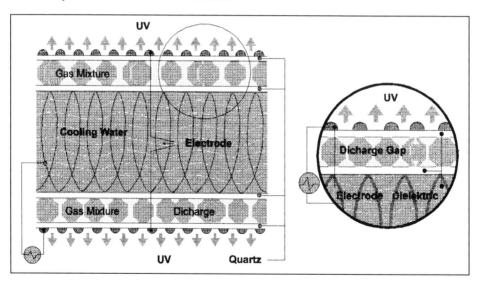

Figure 9 Excimer Lamp and its Configuration (schematic)

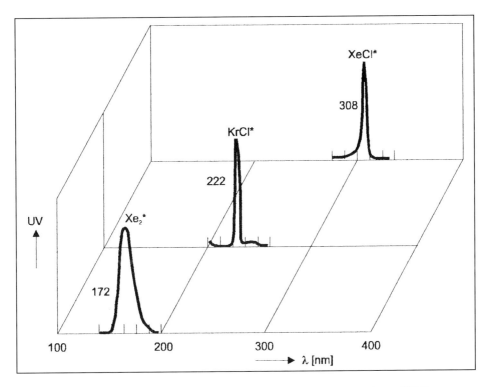

Figure 10 Emission Spectra of the Xe$_2$, KrCl and XeCl Excimer UV Lamps

Other lamp geometries such as flat lamps with a planar geometry and lamps with special reflectors are also offered. An important feature of the excimer lamps is their cold surface. The heat generated by the discharge is completely transferred to the cooling water. There is practically no heat transfer from the excimer lamp to the objects to be irradiated. All lamps are radio frequency (RF) powered, and can be easily controlled and switched. Lamp cooling is accomplished by deionised water. The power supply includes all electrical controls, interlocks and the cooling water supply system.

Table III Specifications of Heraeus Excimer Lamps

Excimer	Peak wavelength (nm)	Specific electrical power (Wcm^{-1})	UV radiant power (Wcm^{-1})	Maximum lamp length (cm)	Main applications
Xe$_2^*$	172	15	(1-1.5)*	175	Physical matting, Photodeposition, Surface treatment, Ozone generation
KrCl	222	50	5	100	UV oxidation, Dry etching, Photochemistry, UV curing, Sterilisation
XeCl	308	50	5	100	UV curing Photochemistry
KrBr**	282				

* estimated, ** in preparation

3. Emission from microwave discharge powered excimer lamps

In 1989 Kumagai and Obara [3] showed that microwave discharge pumping of ArF and KrF excimers can be applied to generate the corresponding incoherent excimer radiation. The experimental set-up used consisted of a 2.45 GHz magnetron as microwave source. The microwaves were fed through a directional coupler, a tuner and a coupling slit into a cylindrical cavity made of a metallic mesh, where they created a strong electromagnetic field in the TE$_{111}$ mode. An electrodeless cylindrical discharge tube containing a fluorine rare gas mixture was placed in the position of the highest electrical field strength. At typical filling gas pressures of 50 to 200 mbar, the UV radiant power of the ArF and KrF excimer lamp reached 29 W and 53 W, respectively. If compared with the microwave power deposited, this corresponded to a UV power efficiency of 4.4 and 8.3% [16, 17].

Based on these results, efforts have been made to develop a compact, low cost excimer lamp system for industrial applications. Using an excimer lamp configuration composed of two 2.45 GHz, 5kW microwave generators, two waveguides, coupling antennae and a microwave cavity (see Figure 11), excimer UV radiation from XeCl and KrCl was generated.

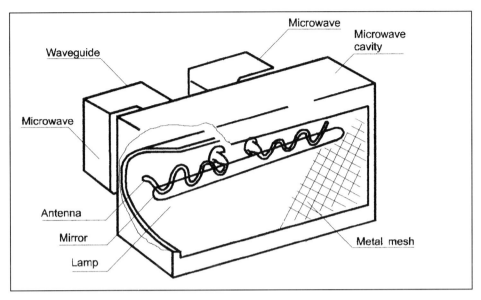

Figure 11 Microwave Discharge Powered Excimer Lamp

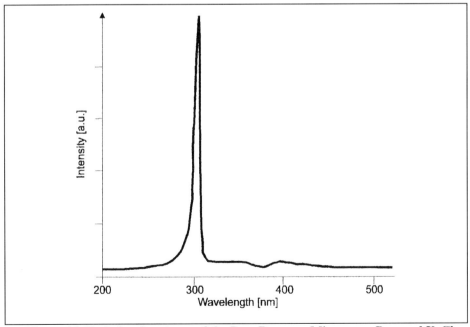

Figure 12 Emission Spectrum of the Low Pressure Microwave Powered XeCl Excimer Lamp

A maximum UV radiant power of 80 W was achieved for XeCl [18]. With a bulb length of 25 cm, this corresponds to the relatively low specific UV output of 3.2 Wcm^{-1}. A narrow 308 nm band with a full width half maximum of about 5 nm was emitted, which did not change its shape up to 10 kW microwave excitation power.

The UV radiant power of such a lamp configuration is limited by the low pressure of the lamp fills. Pressures from 50 to 200 mbar are usually applied. At higher pressures the microwave field is unable to ignite the plasma. Efficient electron capture by the halide fills and a higher collision probability at increasing pressure prevents the development of the electron avalanche.

To overcome the limitation imposed by the microwave ignition conditions, auxilary ignition methods have been developed by Frank, Cekic and Wood from Fusion UV Systems, Inc. (Gaithersburg, USA) [19]. They utilise a high voltage pulse, novel additives and special microwave field management to ignite XeCl or KrCl excimer lamps with filling pressures up to 3.6 bar.

A scheme of the high voltage high frequency system is shown in Figure 13. The ignition supply provides line synchronised 1 µs, 2 MHz pulses, gated on for 300 ms. The pulse peak voltage is 60 kV and peak current about 10 A.

The high voltage pulse, additives providing electron field emission during the high voltage pulse and electron injection close to the maximum microwave electrical field strength, act together to create localised pressure reduction and high electron density.

The microwave field applied drives this region into an avalanche discharge, ignites the bulb, and subsequently heats the plasma to its operation point.

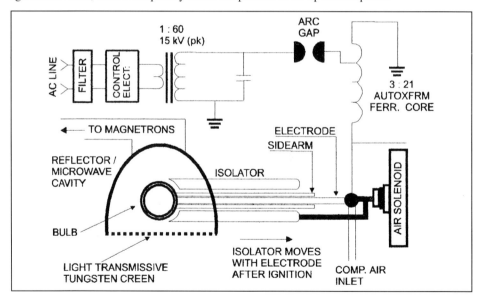

Figure 13 Scheme of the High Voltage High Frequency Ignition System as Used for High Pressure Excimer Lamps

Figure 14 shows the emission spectra of a XeCl and a KrCl high pressure microwave driven excimer lamps. The total UV output (or radiant) power can be obtained by integrating the radiant power density over the wavelength range. For XeCl, about 900 W are emitted in the wavelength range from 280 to 315 nm. Both the KrCl (222 nm) and the Cl_2 excimer transitions (259 nm) are observed in the emission spectrum of the KrCl lamp.

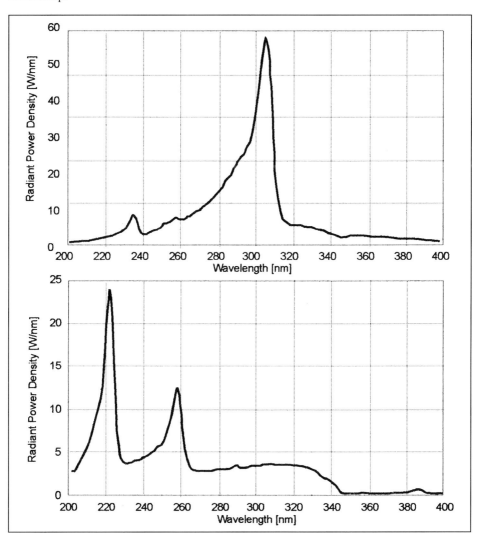

**Figure 14 Radiant Power Density as a Function of Wavelength:
Upper Curve: XeCl, Lower Curve :KrCl and Cl_2**

Knowledge of plasma parameters, such as electron temperature and density is of great importance to the development and optimisation of microwave excited excimer lamps. Using spectroscopic methods of plasma diagnostics, Popovich and Cekic [9] determined the following fundamental plasma parameters of a high pressure microwave excited excimer lamp:

Average electron number density: "3 x "10^{17} cm^{-3}
Average electron temperature: 3600 to 5200 K
Ionisation ratio : 1%
Electron collision rate : 5.5 x10^{12} s^{-1}.

It is important to note that the electron collision rate exeeds the microwave frequency by three orders of magnitude.

In comparison to the plasma conditions found in the microdischarges of a barrier discharge, the average electron density observed in high pressure microwave discharge is more than two orders in magnitude higher.

4. Commercial high pressure microwave excited excimer lamps

In 1996, Fusion UV Systems, Inc. (Gaithersburg, USA) commercialised a 240 Wcm^{-1} XeCl excimer lamp within its VIP (Versatile Irradiance Platform) series. The VIP series is Fusions 10 inch (25.4 cm) UV lamp system with interchangeable bulbs: 308 nm, 535 nm, 396 nm, and aoo to 430 nmAn industrial version of XeCl bulb with enhanced shorter UV radiation is in preparation. The VIP bulbs are high pressure bulbs and are not compatible with other Fusion microwave powered lamp systems.

In contrast to previous Hg based microwave UV lamps, the spectrum of Nobelgas-Halide excimers remains self-similar operating at all power levels. The characteristic radiated energy redistribution of Hg based bulbs is not observed.

Table IV Specifications of Fusion UV Systems Excimer Lamps

Excimer	Peak wavelength (nm)	Specific electrical power (Wcm^{-1})	Radiant power (Wcm^{-1})	Max. lamp length (cm)**	Potential applications
XeCl	308	240	64 (200-400 nm)	25	UV curing
KrCl/Cl$_2$*	222/258	240	33 (200-340 nm)	25	UV oxidation, Surface treatment Dry etching, Photochemistry, UV curing
CuBr***	535	240	33 (530-550)	25	VIS light curing, Laminating thick films

*   in preparation,
** all specifications are for a single lamp system. A larger system can be created by adding more irradiators and lamp units,
*** main transition of the VIP 535 bulb.

It should be pointed out that the UV radiant power of the excimer lamps as specified in Table 4, is considerably higher than that of a 240 Wcm^{-1} medium pressure mercury lamp. Mercury arc lamps reach UV radiant power levels of about 40 Wcm^{-1} in the wavelength range from 200 to 400 nm and about 25 Wcm^{-1} from 200 to 340 nm. This means that electrodeless microwave excited rare gas halide excimer lamps are presently among the most powerful UV sources.

The bulb surface is heated up during operation. Intensive air cooling (11 m^{3}min^{-1}) is essential for stable lamp operation. Water cooling cannot be applied for microwave driven bulbs. Thus, heat transfer to the substrate cannot be completely avoided.

III. EXCIMER LAMPS IN COMPARISON WITH MEDIUM PRESSURE MERCURY LAMPS

In UV curing applications such as curing of printing inks [20], the microwave excited Fusion 308 nm excimer lamp is comparable to a medium pressure mercury lamp:
- very high cure speeds are possible,
- space reqirements are low,

- ozone is generated if curing is performed under air,
- heat is transferred to the substrate or machine,
- substrate decomposition cannot be completely excluded.

A comparison with the mercury lamp also shows that the Fusion 308 nm lamp needs less (lamp) electrical power to reach a comparable cure speed and leads to less substrate decomposition.

The Heraeus 308 nm excimer lamp is not as powerful as the Fusion excimer or a mercury arc lamp, but the following interesting characteristics have been found:
- medium to high cure speed possible,
- no ozone formation, no heat transfer to the substrate or machine,
- no substrate decomposition
- small dimensions,
- simple control.

Medium pressure mercury and Fusion 308 nm excimer lamps are favourably applied as final curing unit if maximum cure speed is needed and temperature insensitive substrates are used.

Heraeus 308 nm lamps can be favorably applied if temperature sensitive substrates have to be treated, interdeck curing is desired and final curing at medium cure speed can be tolerated. However, due to their relatively low UV radiant power, nitrogen inerting has to be used in industrial curing applications.

According to supplier specifications, there are practically no significant differences in the lifetimes of 308 nm excimer and mercury arc lamps. Usually lamp lifetimes of 2000 hours are rated. Due to the interaction of 172 nm photons with quartz impurities and gas/wall reactions, the lifetime of Xe_2 excimer lamps is in the order of several hundred hours.

Figure 15 summarises the range of excimer wavelengths commercially available and compares these with bond energies and the spectral distribution of a medium pressure mercury lamp. It can be clearly seen that excimer radiation can be utilised to excite and dissociate nearly all types of molecules. In addition to transitions at 259 nm (Cl_2), 282 nm (XeBr) and 308 nm (XeCl), which are within the wavelength range of mercury arc emission lines, high energy photons are provided at 126 nm (Ar_2), 172 nm (Xe_2) and 222 nm (KrCl). Using 172 or 222 nm radiation, direct photolysis of e.g. acrylates becomes possible. Radicals generated from excited states are able to directly initiate acrylate polymerisation. It is known from electron pulse radiolysis and laser photolysis studies [21] that acrylate radical formation from excited states is less efficient than that from ionic precursors, but high power sources of high energy photons offer an interesting potential for the technical application of photoinitiator-free curing of acrylates.

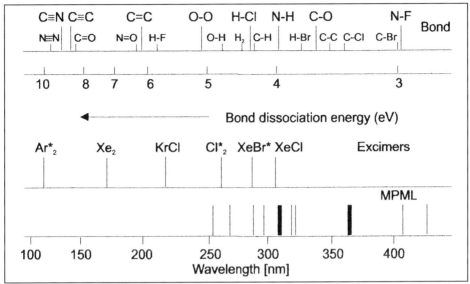

* excimer source in preparation

Figure 15 Wavelengths of Commercially Available Excimer Sources Compared with Bond Dissociation Energies and Transitions from a Medium Pressure Mercury Arc Lamp (MPML)

Excimer emission of 259, 283 and 308 nm in particular fits well into the optical absorption spectra of many photoinitiators. Large extinction coefficients of the photoinitiator at the excimer emission wavelength are an important prerequisite of high radical formation and polymerisation rates (see the formula given for the polymerisation rate v_p in Figure 10, Chapter I). Figure 15 illustrates this for two frequently used photoinitiators: IC 369 and IC 651. It can easily be seen that the second absorption maximun of IC 369 is very close to 308 nm. The extinction coefficient ε_{308} is measured to be 15.8 x $10^4 dm^3 mol^{-1} cm^{-1}$. Using IC 651, the 308 nm emission does not match photoinitiator absorption very well. This is reflected by real-time FTIR (Fourier transform infrared) measurements of the time profile of the double bond conversion in the system photoinitator/TPGDA. Figure 16 shows that, when 308 nm excitation is used, conversion is faster in the case of IC 369, and the inhibition time which is needed to deplete the oxygen concentration in the cured layer is lower than for IC 651. However, the picture would change if e.g. 283 or 259 nm excimer radiation is used.

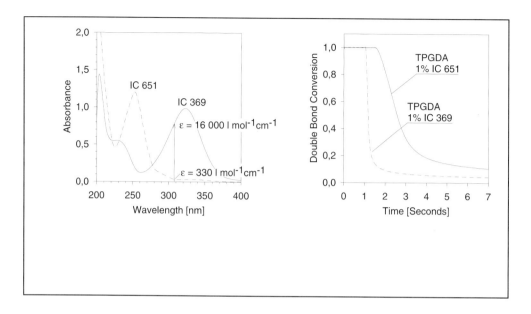

Figure 16 Double Bond Conversion as a Function of Time: Systems IC 369/TPGDA and IC651/TPGDA Irradiated with 308 nm at a Irradiance of 40 mWcm^{-2}

With further advances in technical development, incoherent excimer radiation will become available as a basic tool for UV curing, comparable in its importance, but superior in many respects, to polychromatic radiation from a medium pressure mercury arc discharge.

IV. REFERENCES

1. B. Eliasson, U. Kogelschatz: Appl. Phys. B 46, p. 299 (1988)
2. Ch.K. Rhodes: Excimer Lasers, Topics Applied Physics, Vol. 30 Springer, Berlin, Heidelberg (1984)
2. H. Kumagai, M. Obara: Appl. Phys. Letters 54, p. 2619 (1989) and 55, p. 1583 (1989)
4. U. Kogelschatz: Pure & Appl.Chem. Vol 62, p. 1667 (1990)
5. U. Kogelschatz: Proceedings Thenth Int. Conf. Gas Discharges and Their Applications, Vol.II, p.972 (1992)
6. I.W. Boyd, Y.Y. Zhang: Nucl. Instr. And Meth. In Phys. Res. B 121, p. 349 (1997)
7. B. Gellert, U. Kogelschatz: Appl. Phys. B 52, p. 14 (1991)
8. K.L. Kompa, H. Walther (eds.): High Power Lasers and Applications, Springer Ser. Opt. Sci., Springer, Berlin, Heidelberg (1987)
9. S. Popovic, M. Cekic : ICOPS/IEEE Conference, San Diego (1997)
10. D.J. Eckstrom, H.H. Nakano, D.C. Lorents, T. Rothem, J.A. Betts, M.E. Lainhart, D.A. Dakin, J.E. Maenchen: J. Appl. Phys. 64, p. 1691 (1988)
11. U. Kogelschatz in : Nato ASI Series Vol. G 34, Non-Thermal Plasma Techniques for Pollution Control, edited by M. Penetrante and S.E. Schultheis, Springer Berlin, Heidelberg, p.339 (1993)
12. H. Esrom, H. Scheytt, R. Mehnert, C. von Sonntag in: Nato ASI Series Vol. G 34, Non-Thermal Plasma Techniques for Pollution Control, edited by M. Penetrante and S.E. Schultheis, Springer Berlin, Heidelberg, p.91 (1993)
13. H. Esrom, U. Kogelschatz: Thin Solid Films 218, p. 231 (1992)
14. U. Kogelschatz: Applied Surface Science 54, p. 410 (1992)
15. R.S. Nohr, J.G. MacDonald, Radiat. Phys. Chem. 46, p. 983 (1995)
16. H. Kumagai, M. Obara: Appl. Phys. Lett. 54, p. 2619 (1989)
17. H. Kumagai, M. Obara: Appl. Phys. Lett. 55, p. 1583 (1989)
18. M. Kitamura, K. Mitsuka, H. Sato: Applied Surface Science 79/80, p. 507 (1994)
19. J.D. Frank, M. Cekic, C.H. Wood: Proccedings of 32nd Microwave Power Symposium, Ottawa, Canada, p. 60 (1997)
20. R. Mehnert, U. Decker, W. Arnold: Papier+Kunsstoffverarbeiter 9-97, p. 57 (1997)
21. R. Mehnert, W. Knolle: Proceedings RadTech Europe`95, Vol. Academic Day, p. 135 (1995)

CHAPTER V

DOSIMETRY FOR EB AND UV CURING

I. INTRODUCTION

Dosimetry is a branch of physical or physico-chemical sciences dealing with the measurements of quantities characterising radiation fields and radiation "doses" (exact definitions of quantities will be given later). The term dosimetry, despite being general, today mainly relates to ionizing radiation (gamma radiation, electron beams, etc.). For the measurement of UV radiation and visible light - photodosimetry – the terms radiometry and actinometry are currently used, related to the measurement of radiant energy (with physical devices) and the number of photons (usually with chemical systems), respectively.

The primary need for dosimetry in any process where radiation is used is quite straightforward. Often it is also used in other irradiation conditions, e.g. radiation intensity. If an attained property of the processed material or the product is related to a certain radiation dose, it is easy to reproduce the procedure. Dosimetry can thus become a methodology which should be applied to ensure that a radiation process meets a given specification and it also enables transfer of technology - developed usually in a research laboratory - to a pilot scale and finally to a fully commercial level. In this respect dosimetry is also indispensable for optimizing the overall technological process and its parameters, especially regarding the efficient utilization of radiation and homogeneous irradiation of material. As for the production phase itself, dosimetry plays an important role in routine quality control and in some cases reliable measurement of specified radiation dose can even be substituted for the more tedious testing of end-point product properties.

Despite the fact that UV curing is today far more widely used than EB curing, dosimetry - as far as the industrial applications are concerned - is apparently much better established in the case of electron beams [1], where even the stage of standardisation has already been achieved [2].

As for the interactions of radiation with matter, there is a fundamental difference between ionizing radiation (electron beams) and UV radiation, and in the case of UV radiation, unlike ionizing radiation, there are effects which are specific with respect to the properties of irradiated material. This explains why concepts of dosimetry are different in both cases and will thus be discussed separately in the next paragraphs. It covers the measurement of energy or photons of a certain energy delivered to radiation processed material.

II. LOW-ENERGY ELECTRON BEAM DOSIMETRY

1. Interactions of electrons with matter and dose

When an electron beam enters a material (accelerator exit window, air gap and the material to be irradiated) the electron energy - originally given by the acceleration voltage - is substantially altered. As a first approximation, electrons may be considered to lose their energy and slow down *continuously*, namely in a very large number of interactions with only small energy losses.

Electrons, as other charged particles, transfer their energy to the material through which they pass in two types of interaction [3,4]:

- in *collisions* with atomic electrons, resulting in material ionization and excitation,
- in interactions with atomic nuclei leading to the emission of X-ray photons. This is usually called *bremsstrahlung* (electromagnetic radiation emitted charged particle changes velocity).

For low-energy electrons only collision interactions are significant.
The most important terms and quantities which are relevant for electron-beam dosimetry are summarized in Table I. Definitions of quantities follow the recommendations of the International Commission on Radiation Units and Measurements (1980) [5].

Absorbed dose D, sometimes simply referred to as dose, is defined as the mean incremental energy $d\varepsilon$ imparted to matter of incremental mass dm. Although this definition, strictly speaking, gives the absorbed dose at a point in radiation-absorbing material, it is generally averaged over a finite mass of a given material, with the absorbed dose being read by a calibrated dosimeter. The increment of the absorbed dose in the time interval dt is *the absorbed dose rate*, $\dot{D}$. Often the dose rate is simply defined as the dose absorbed per unit time, which is acceptable in the case of a continuous radiation; however, in the case of a pulsed beam one has to distinguish between an average dose rate and dose rate in a pulse, which may have a significant effect on radiation chemical processes in irradiated material.

As the absorbed dose is the energy absorbed per unit mass, it can be measured directly via the heat produced by the absorption of radiation, i.e. by calorimetry, which is the primary absolute dosimetric method.

Table I Main Quantities Relevant to EB Dosimetry

Quantity	Symbol	Definition	Unit
Absorbed dose	D	$d\varepsilon/dm$	Gy (Gray) = J kg^{-1} (1 rad = 0.01 Gy)
Dose rate	$\dot{D}$	dD/dt	Gy s^{-1}
Linear stopping power	S	dE/dl	MeV cm^{-1} or J m^{-1}
Mass stopping power	S/ρ	$1/\rho\,(dE/dl)$	MeV cm^2 g^{-1} or J m^2 kg^{-1}
Practical range	R_p		g cm^{-2} or kg m^{-2}

The energy loss of the electron per unit path length is called *linear stopping power S* (according to the type of the interaction either collision, radiation or total stopping power). Dividing the linear stopping power by the density ρ of the material, the *mass stopping power* S/ρ is obtained. The mass collision stopping power is a function of electron energy and material composition. The values for some materials at various electron energies are given in Table II [6]. They can also be calculated for other materials (element, compound, mixture) by a convenient procedure developed by Seltzer and Berger [7,8].

The mass collision stopping power is useful for comparing absorbed doses in different materials: if the number and energy of electrons passing through them is identical, then the ratio of electron energy lost per unit mass, i.e. of absorbed doses, will be equal to the corresponding *stopping power ratio*. Under these conditions, the dose D_m in another material can also be calculated from the dose D_d measured by a dosimeter:

The mass stopping power ratio is usually only a slowly varying function of energy,

$$D_m = D_d \cdot (S/\rho)_m / (S/\rho)_d$$

$(S/\rho)_m$ = mass stopping power of the material
$(S/\rho)_d$ = mass stopping power of the dosimeter

therefore the actual energy does not need to be known with a great accuracy.

Table II Mass Collision Stopping Power [MeV cm^2 g^{-1}]

Material	Electron energy [keV]				
	100	150	200	300	400
Water	4.12	3.24	2.79	2.36	2.15
Graphite	3.67	2.88	2.48	2.08	1.89
Aluminium	3.18	2.51	2.17	1.84	1.68
Titanium	2.87	2.27	1.97	1.67	1.53
Polyamide (nylon 6)	4.15	3.26	2.81	2.37	2.16
Polyethylene	4.38	3.44	2.97	2.50	2.27
Polyethylene terephthalate	3.83	3.02	2.60	2.20	2.00
Polymethyl methacrylate	4.01	3.15	2.72	2.29	2.09
Polystyrene	4.03	3.17	2.74	2.31	2.10
Teflon	3.42	2.70	2.33	1.97	1.80
Polyvinyl chloride	3.60	2.84	2.46	2.08	2.90
Film emulsion	2.68	2.14	1.86	1.59	1.45

When an electron beam impinges on a slab of material, a fraction η of electrons will be *backscattered* due to nuclear elastic interactions. This effect is particularly significant at low electron energies and increases with increasing material thickness and atomic number. In Figure 1 backscatter coefficients for electrons incident on targets from different materials are plotted as a function of the incident energy [6]. Electron backscattering has to be considered from the point of view of the backing material when thin layers (e.g. foils) are irradiated with low-energy electrons.

Electron scattering is also responsible for typical electron *depth-dose distribution* in irradiated material (variation of the absorbed dose with the depth from the incident surface), namely the dose increase at shallow depths. Depth-dose curves for several electron energies - based on Monte Carlo calculations - are shown in Figure 2 [2]; at very low energies increasing parts of the curves are missing, as a significant portion of the electron energy is already absorbed in the accelerator exit window and air gap between the window and material. By extrapolating the straight portion of the descending part of a depth-dose curve to the axis, a quantity commonly called *practical range* (or *extrapolated range*) is obtained. This parameter has no physical significance, however it adequately characterises the penetrating power of electrons. Figure 2 indicates that the practical range of the electrons with energies 0.1-0.3 MeV is of the order of 10^{-2} g cm^{-2} only.

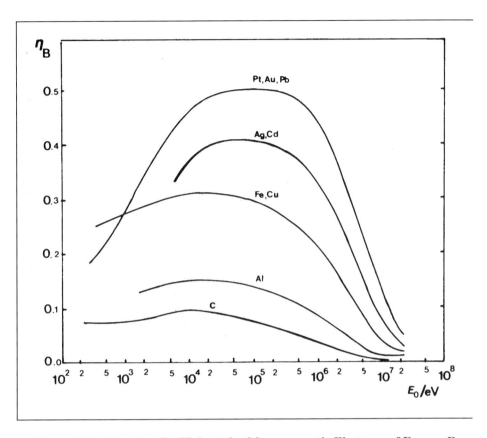

Figure 1 Backscatter Coefficients for Monoenergetic Electrons of Energy E_0 Perpendicularly Incident on a Semi-Infinite Medium [6]

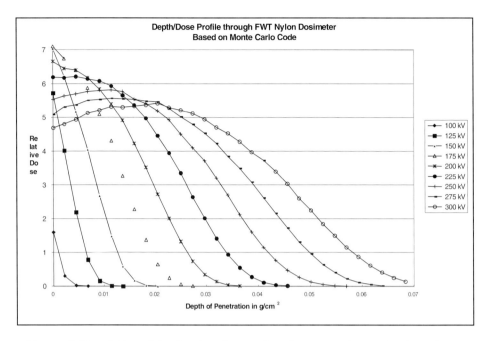

Figure 2 Depth-Dose Distribution Curves for Electrons of Energies from 0.1 to 0.3 MeV in Nylon (0.5 mil Ti Window, 0.5 in. Air Gap) [2]

2. Dosimetry with thin-film dosimeters

A *dosimeter* can be defined as a device which, when irradiated, exhibits a quantifiable change in some property which can be related to the absorbed dose using appropriate analytical instrumentation and techniques. A *dosimetry system* is a system used for determining the absorbed dose, consisting of dosimeters, measurement instruments and their associate reference standards and procedures for using the system.

Owing to the limited penetration of low-energy electron beams and the narrow air gaps in self-shielded irradiators, *thin-film dosimeters* are practically the only dosimeters which can be used in routine practice. Thin-film dosimeters are both undyed and dyed plastics which, when irradiated, exhibit a change in optical absorbance, proportional to the absorbed dose. To exclude errors due to possible thickness variations in individual dosimeters, the *dosimeter response* is usually expressed as the radiation-induced absorbance change divided by the dosimeter thickness. Since the dosimeter averages the absorbed dose over its thickness, the lower the beam energy, the smaller the dosimeter thickness should be, in order to obtain an appropriate and meaningful spatial dose resolution.

(i) Types, Properties and Evaluation of Dosimeters

The most common materials which are suitable for EB dosimetry are summarised in Table III, whereas Table IV lists some dosimeter suppliers.

Table III Optical Absorption Thin-Film Dosimeters for EB Dosimetry

Material (Type)	Film Thickness [μm]	Analysis Wavelength [nm]	Dose Range [kGy]	Ref.
Radiochromic films: Leucocyanides or leucomethoxides of triphenylmethane dyes				
in nylon (FWT-60)	50 or 10	600 or 510	0.5-200	(9-11)
in polychlorostyrene (FWT-70)	50	630 or 430	1-300	(9-11)
in polyvinyl butyral (Risø B3 or B4)	22	553	1-200	(12)
GafChromic film: radiochromic microcrystalline layer on a 100 μm polyester support	6 (sensor)	650 or 400	0.1-50	(13-15)
Cellulose triacetate	38 or 125	280	5-300	(16,17)
Dyed cellulose diacetate (DY-4.2)	130	390-450	10-500	(16,18)
Dyed (blue) cellophane	20-30	650	5-300	(10)

Table IV Some Suppliers of Film Dosimeters for EB Dosimetry

Dosimeter Type	Supplier
FWT radiochromic films	Far West Technology, 330 South Kellog Goleta, California 93117, USA
Risø B3 radiochromic films	Risø National Laboratory, Roskilde, Denmark
HD-810 GafChromic Dosimetry Media	Victoreen, Inc., Nuclear Associates Div., 100 Voice Rd., Carle Place, NY 11514-1593
FTR-125 undyed cellulose triacetate	Fuji Photo Film Co., 2-26-30 Nishiazabu, Minato-ku, Tokyo, 106 Japan
DY-4.2 dyed cellulose diacetate	Department of Materials and Nuclear Engineering, University of Maryland, College Park, MD 20742-2111, USA
Polymer-alanine film dosimeter	Gamma-Service Produktbestrahlung GmbH, Juri-Gagarin-Strasse 15, D-01454 Radeberg, Germany

The response of any dosimeter can generally be influenced by the following factors:
- the way of delivering the dose, namely the dose rate and possible dose fractionation,
- environmental factors, namely temperature, humidity, oxygen content and light,
- stability after irradiation.

These effects should always be checked - even for each material batch - and if serious, it is advisable to calibrate dosimeters under conditions which are as close as possible to those applied in routine use, otherwise corresponding corrections have to be made.

Radiochromic film dosimeters show no dose-rate dependence (at least in a certain dose range, usually up to about 50 kGy) and that is why they are very suitable for EB dosimetry. They are quite sensitive to UV light (< 370 nm), therefore they also have to be protected from ambient light. After a short-term, irradiation with electron beam absorbance somewhat slowly increases for several hours, which can be eliminated by a post-irradiation heat treatment (5 min at 60 °C). Both FWT and Risø films show combined temperature-humidity dependence [19-21]; it is advisable to avoid using them at extreme relative humidities < 30% or >70%, as well as at temperatures above about 60 °C.

GafChromic films also show slow additional absorbance increase after irradiation, and it is recommended to read them about 24 hours later. The effects of humidity on the response are not very pronounced [19].

For undyed cellulose triacetate it is recommended that the absorbance should be measured after it has become stable, i.e. approximately 2 h after irradiation. At high dose rates in electron beams, the film does not show significant temperature and humidity dependence. The film response at low dose rates - typical for gamma-ray sources - is by about 20% higher than in electron beams; this response change occurs in a dose-rate range of 10^4 to 10^5 Gy/h and can be eliminated if the film is irradiated in an inert atmosphere [16,17].

Blue cellophane (containing diazo dye) which was widely used in the past was supplied by several companies in the USA, Japan and Germany; most of them, however, apparently stopped production some years ago. The advantages of this material were ease of use and availability of rolls or long strips, but there were also problems due to the non-uniform distribution of the dye and striations in the cellophane core material, as well as fairly pronounced effects of temperature and humidity [18].

The absorbance of film dosimeters can be evaluated by conventional spectrophotometry, by spatial-scanning spectrophotometry (for such purposes a gel-scanner which is supplied with many spectrophotometers can be conveniently utilized), or - when very high resolution is required - also by microdensitometry. There are also single-purpose absorbance measuring devices on the market. The Far West Technology Reader (FWT-92) is designed to measure FWT radiochromic films and can be modified for reading GafChromic films. Nuclear Associates (Victoreen) supplies a Radiochromic Densitometer with a film transport system for use with GafChromic films. A photometer NHV FDR-01 (supplied by NHV America Inc.) with scanning equipment can be applied to undyed cellulose triacetate dosimeters. A flexible instrument (comprising spectrophotometer, thickness gauge and software) for evaluating virtually any optical absorption dosimeter has been developed by AERIAL (Schiltigheim, France). A special high-resolution micro-densitometer (Model CMR-604) for GafChromic films is offered by PeC Photoelectron Corporation (USA).

Dose measurement using a special type of dosimeter, i.e. a polymer-analine film, which is also listed in Table IV, is based on the quantitative determination of stable free radicals in irradiated microcrystalline aminoacid analine, by electron paramagnetic resonance (EPR) spectroscopy [22,23]. The sensitive layer (thickness from 5 to 100 µm), composed of analine and a binder, is deposited on a polyester support (thickness from 23 to 100 µm). Depending on the sensitive layer thickness, doses in a wide range from 0.2 to 500 kGy can be measured. Analine-based dosimeters generally show excellent dosimetric properties, namely dose-rate independent response, response long-term stability and low temperature and humidity dependence. The main disadvantage regarding their routine use is the need for an EPR spectrometer. However, less expensive instruments designed for dosimetry purposes are on the market, e.g. the Bruker EMS 104 EPR Analyzer.

(ii) Calibration of Dosimeters

The film dosimeter response is best calibrated in electron beam against a calorimeter, and this method is well established for beams in the energy range 4 - 12 MeV [1,24], as well as 1-3 MeV [25]. A typical response curve of the Risø B3 radiochromic film as obtained by calibration in a 10-MeV electron beam of the Linac "ELEKTRONIKA" against graphite calorimeter is shown in Figure 3.

A simple calorimeter with a totally-absorbing graphite body designed for measuring energy deposition in materials and calibrating the response of dosimeters under dynamic conditions in the beams of low-energy accelerators has been successfully applied to a 400-keV electron accelerator as well [26]. The film calibration based on the irradiation of the total-absorption calorimeter and total-absorption stack of film dosimeter to be calibrated - under identical conditions - is shown schematically in Figure 4. The simple expression given for calculating the calibration factor is only valid if the dose dependence of the radiation-induced absorbance is linear. In other cases the relative dose-response function has to be determined, and the response of films in the stack corrected by this function.

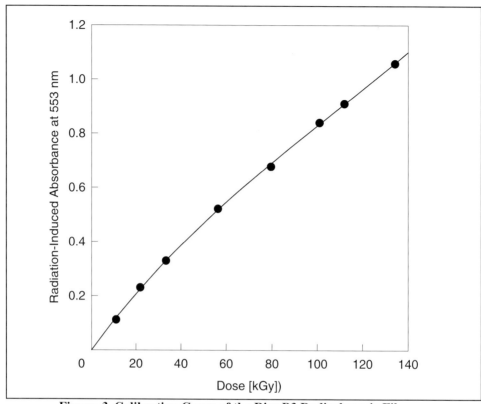

Figure 3 Calibration Curve of the Risø B3 Radiochromic Film

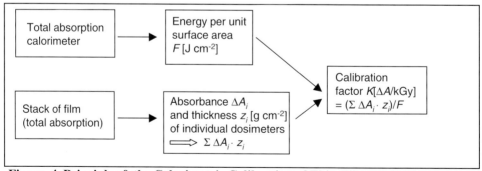

Figure 4 Principle of the Calorimetric Calibration of Thin-Film Dosimeters in a Low-Energy Electron Beam [26]

The calibration of a user's routine dosimetry system - when *traceability* to a national standards laboratory is required - can be established in three different ways:
- by irradiating dosimeters at a calibration laboratory and measuring their response at the user's laboratory,
- by irradiating dosimeters in the user's in-house calibration facility, which has been calibrated via transfer dosimeters from a calibration laboratory,
- by irradiating dosimeters directly in the user's production facility, together with the transfer dosimeters which are provided and read by the calibration laboratory.

Such calibrations are often carried out by using ^{60}Co-gamma-ray sources, which of course requires the dosimeter to be dose-rate independent.

(iii) Applications of Thin - Film Dosimeters

Whenever thin-film dosimeters are used for dose measurements in low-energy electron beams, the backscattered electrons from the beam stop have to be prevented from affecting the dose measurement. For that reason a backing material (similar in composition to the dosimeter) of a thickness equal to the electron practical range should be used. Using the film dosimeters the following measurements are commonly carried out [2,27].

- The *surface area rate* (or *processing coefficient*) K, relating the area irradiated per unit time to the beam current and the absorbed dose, is determined by measurement of the surface dose (over the dynamic operating voltage of the accelerator) at several beam current levels [2].

Area processing rate [m^2 min^{-1}] = $W_b \cdot V = K \cdot I / D$

K = surface area rate [kGy m^2 mA^{-1} min^{-1}]
D = dose [kGy]
W_b = beam width [m]
V = conveyor speed [m min^{-1}]
I = beam current [mA]

- The *cure yield* k, calculated in terms of the average absorbed dose D delivered to the layer of interest times the conveyor speed per unit effective beam current, can be estimated by measuring the dose over the range of beam currents at different speeds[27].

> $k = D\,V - D$ dosimetry for EB and UV Curing **V / I**
> k = cure yield [kGy m mA^{-1} min^{-1}]
> D = dose [kGy]
> V = conveyor speed [m min^{-1}]
> I = beam current [mA]

- *The dose uniformity across the width of the beam* should be measured by placing dosimeters at intervals of about 2.5 cm across the width of the beam (or using a long strip of film) and the observed variation in dose should fall within the range specified by the accelerator manufacturer [2].

- *The depth-dose distribution* can be measured by irradiating the dosimeter stack with a thickness slightly greater than the practical range at the energy of interest. By evaluating individual dosimeters a depth-dose profile is obtained. Experimental results can be conveniently compared with the results of relatively simple calculations carried out by using the computing code EDMULT (Energy Deposition in MULTilayer absorber), which is based on a semi-empirical model [28,29]. The code is valid for incident energies from 0.1 to 20 MeV and for absorbers consisting of up to six slabs of different materials of effective atomic numbers from about 5.6 (polystyrene) to 82 (Pb) [28,29]. Thus the accelerator exit window, the gap between the window and the cured material, the cured material itself, as well as the backing material, can be taken into account. An example where the depth-dose distribution in polyethylene terephthalate (PET) was measured by using a thin (22 µm) Risø B3 radiochromic film is shown in Figure 5. A stepped array of eight thin PET films (23 µm each) was placed upon a single strip of the radiochromic film, which had as a backing material a 400-µm thick PET layer. After irradiation, the "stepped absorbance" of the radiochromic film strip was evaluated by using a gel scanner, and doses in different depths in PET were thus obtained. Figure 5 shows also the results of calculations performed by using EDMULT.

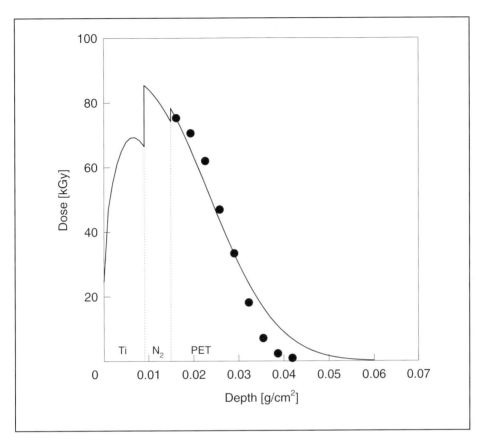

Figure 5 Comparison of Measured (Points) and Calculated (Curve) Depth-Dose Distributions in PET for 180-keV Electron Beam (LEA-1 Accelerator) Passing First Through a 20-μm Ti Window and a 5-cm Nitrogen Gap (Normalised at the Maximum Measured Dose)

III. ULTRAVIOLET RADIATION DOSIMETRY

1. Basic terms, quantities and units

Definitions of quantities - which appear to be most important in UV-radiation dosimetry - are taken from the IUPAC Recommendations 1996 [30]. They are summarized, together with the symbols and units, in Table V.

Radiometric quantities (related to the energy of electromagnetic radiation):
Radiant power (P) - Same as *radiant (energy) flux (Φ)*. Power emitted, transferred or received as radiation.

Radiant energy (Q) - The total energy emitted, transferred or received as radiation of all wavelengths in a defined period of time.

Spectral radiant power (P_λ) - The *radiant power* at *wavelength λ* per unit wavelength interval.

Irradiance (E) - The *radiant flux* or *radiant power* of all wavelengths, incident on an infinitesimal element of surface containing the point under consideration divided by the area of the element. For a parallel and perpendicularly incident beam not scattered or reflected by the target or its surroundings, the *fluence rate (E_0)* is the equivalent term.

Spectral irradiance (E_λ) - Irradiance at *wavelength λ* per unit wavelength interval.

Radiant exposure (H) - The *irradiance* integrated over the time of irradiation.

Photonic quantities

Einstein - one mole of photons.

Photon flow (Φ_p) - The number of photons (quanta) per unit time. Alternatively, the term can be used with the amount of photons (mol or *Einstein*) per unit time.

Spectral photon flow ($\Phi_{p\lambda}$) - The *photon flow* at *wavelength λ* per unit wavelength interval. Alternatively, the term can be used with the amount of photons (mol or *Einstein*).

Photon irradiance (E_p) or *photon flux* - The *photon flow*, incident on an infinitesimal element of surface containing the point under consideration divided by the area of that element. For a parallel and perpendicularly incident beam not scattered or reflected by the target or its surroundings, the photon *fluence rate (E_p^0)* is an equivalent term.

Spectral photon flux (photon irradiance) ($E_{p\lambda}$) - The *photon irradiance* at *wavelength λ* per unit wavelength interval. Alternatively, the term can be used with the amount of photons (mol or *Einstein*).

Photon exposure (H_p) - The *photon irradiance* integrated over the time of irradiation. Alternatively, the term can be used with the amount of photons (mol or *Einstein*). For a parallel and perpendicularly incident beam not scattered or reflected by the target or its surroundings, photon *fluence (H_p^0)* is an equivalent term.

Quantum yield (Φ) - The number of defined events which occur per photon absorbed by the system. For a *photochemical reaction* Φ equals amount of reactant consumed or product formed divided by the amount of photons absorbed.

Dose (UV-dose)

The energy or amount of photons absorbed per unit area or unit volume by an irradiated object during a particular exposure time. The term is used in the sense of *fluence*, i.e., the energy or amount of photons per unit area or unit volume received by an irradiated object during a particular exposure time.

Intensity - Traditional term for *photon flux, fluence rate, irradiance*, or *radiant power (radiant flux)*. In terms of an object exposed to radiation, the term should now be used for qualitative descriptions only.

Photon energy
is given by Planck's equation:

$$E = hc/\lambda = 1.986 \cdot 10^{-16} \cdot 1/\lambda \quad [\text{J photon}^{-1}]$$

h = Planck's constant
c = speed of light
λ = wavelength of radiation [nm]

Table V Quantities and Units Related to UV Radiation Dosimetry

Quantity	Symbol	SI Unit	Common unit
Dose		J m^{-2}, $\text{mol m}^{-2\,a}$ or J m^{-3}, $\text{mol m}^{-3\,a}$	
Irradiance	E	W m^{-2}	
Photon exposure	H_p	m^{-2}, $\text{mol m}^{-2\,a}$	
Photon flow	Φ_p	s^{-1}, $\text{mol s}^{-1\,a}$	
Photon irradiance Photon flux	E_p	$\text{m}^{-2}\text{ s}^{-1}$, $\text{mol m}^{-2}\text{ s}^{-1\,a}$	
Radiant energy	Q	J	
Radiant power	P	W	
Radiant exposure	H	J m^{-2}	
Spectral irradiance	E_λ	W m^{-3}	$\text{W m}^{-2}\text{ nm}^{-1}$
Spectral photon flow	$\Phi_{p\lambda}$	$\text{s}^{-1}\text{ m}^{-1}$ $\text{mol}^{-1}\text{ s}^{-1}\text{ m}^{-1\,a}$	$\text{s}^{-1}\text{ nm}^{-1}$ $\text{mol s}^{-1}\text{ nm}^{-1\,a}$
Spectral photon flux Photon irradiance	$E_{p\lambda}$	$\text{s}^{-1}\text{ m}^{-3}$ $\text{mol s}^{-1}\text{ m}^{-3\,a}$	$\text{s}^{-1}\text{ m}^{-2}\text{ nm}^{-1}$ $\text{mol s}^{-1}\text{ m}^{-2}\text{ nm}^{-1\,d}$
Spectral radiant power	P_λ	W m^{-1}	W nm^{-1}

a) If amount of photons is used.

Actinometer
By definition [30] an actinometer is a chemical system or physical device which determines the number of photons in a beam either integrally or per unit of time, and this name is commonly applied to devices used in the ultraviolet and visible wavelength ranges. In a chemical actinometer the photochemical change can be directly related to

the number of photons absorbed while physical devices give a reading that can be correlated to the number of photons detected.

2. Radiometers

The quantitative physical detectors of radiation - commonly called radiometers - convert the radiant energy to an electrical signal and can basically be divided into two classes: thermal detectors and photon detectors [31].

Thermal detectors

In thermal detectors, the incident photon energy is converted into heat which is then measured. They include thermocouples or thermopiles, bolometers and pyroelectric detectors, when temperature change is measured via the thermoelectric effect, via the change in electric resistance and via the change in capacitance, respectively. The response of thermal detectors is independent of the wavelength, and these detectors can measure radiant energy or power in any spectral region. If the wavelength (for monochromatic radiation) or intensity vs. wavelength distribution (for polychromatic radiation) is known, photon flux can also be determined. Thermal detectors can be calibrated in absolute manner by electrical substitution and can serve as standards.

Photon detectors

Photon detectors are based on the photoelectric effect. In contrast to thermal detectors, their response is spectral-dependent. There are three types of photon detectors: photomultipliers based on photoemission (external photoelectric effect) and photovoltaic cells (photodiodes) and photoconductors, both based on the internal photoelectric effect in the semiconductor.

Spectroradiometers equipped with a photon detector and a monochromator enable measurement of the spectrum or intensity of each component of a radiation source.
Most of the radiometers designed for routine UV light measurement operate on the basis of photodiodes.

Some commercially available radiometers - which can be used for controlling and monitoring UV curing systems - are listed in Table VI. Table VII gives some low-profile self-contained instruments for on-line measurement (e.g. in UV-curing units) of the integral UV dose delivered to the cured material, peak irradiance, as well as the irradiance/temperature profiles.

Table VI Some Radiometers for UV Light Measurement

Radiometer	Spectral Range [nm]	Measuring Range [a]	Producer and/or Supplier
Sola-Scope Spectroradiometer	240-425 (1 nm steps)	min $10 nW/cm^2/nm$ max. $> 10 W/cm^2$	Solatell, 4D Controls Ltd., Redruth, Cornwall, UK
SpotCure UV Intensity Meter	250-260 280-320 320-390 395-445	0-19.99 W/cm^2	EIT, Inc., Sterling, VA, USA
UV Curing Radiometer System IL1745 System IL1445	250-400 250-400	1 nW - 1 W/cm^2 150 nW-0.5 W/cm^2	International Light Inc., Newburyport, MA, USA
UVX Radiometer	250-290 280-340 335-380 (three sensors)	0-20 mW/cm^2 0-200 $\mu W/cm^2$ 0-2000 $\mu W/cm^2$ 0-200 mW/cm^2	UVP Inc., Upland, CA, USA
J-225 Meter J-221 Meter	220-280 300-400		
URADES 4	225, 310 and 254 (single sensor)	0-2 mW/cm^2 0-20 mW/cm^2 0-200 mW/cm^2 0-2000 mW/cm^2	G.E.R.U.S. mbH Berlin, Germany

[a] irradiance or spectral irradiance

Table VII Some Self-Contained Low-Profile Devices for On-Line UV Light Measurement

Radiometer	Spectral Range [nm]	Measuring Range [a]	Producer and/or Supplier
UVIRAD	320-390	0-9999 mJ/cm^2 (0.1-100 mW/cm^2) [b]	EIT Inc., Sterling, VA, USA
UVICURE Plus	320-390	0-250 J/cm^2 peak int. 0-5 W/cm^2 (0.005-5 W/cm^2) [b]	
UV Power Puck	250-260 280-320 320-390 395-445	0-250 J/cm^2 peak int. 0-5 W/cm^2 or 0-10 W/cm^2	
UVIMAP	250-260 280-320 320-390 395-445	5-100 mW/cm^2 or 0.1-5 W/cm^2, 0 - 400 °C	
MicroCure MC-2 MC-10	320-390	0-9999 J/cm^2 (up to 2 W/cm^2) [b] (up to 10 W/cm^2) [b]	
LIGHT BUG IL290 IL390B	205-345 250-400	0.1-20 000 mJ/cm^2 (up to 2.5 W/cm^2) [b]	International Light Inc., Newburyport, MA, USA
CON-TROL-CURE Compact Radiometer	250-410	10-20 000 mJ/cm^2 (0.0001-2.5 W/cm^2) [b]	UV Process Supply, Chicago, IL, USA

[a] UV dose, irradiance, temperature
[b] irradiance operating range

3. Chemical actinometers

For practical reasons, among different chemical systems that have been proposed for the integration of incident light [32], only liquid-phase solution actinometers appear to be commendable for use.

Actinometry of basically monochromatic radiation can be performed under conditions of full light absorption (absorbance of the actinometric solution A_λ in the irradiation cell at the irradiation wavelength λ is ≥ 2 during entire irradiation) or partial light absorption ($A_\lambda < 2$) in an actinometric solution. The latter can be carried out either with low solute conversion (changes in A_λ during irradiation are less than 10%) or with high conversion. Chemical actinometry then requires the experimental determination of the number of actinometric molecules Δn_{Ac} which reacted during a given irradiation period (number of molecules of solute consumed or of product formed); in the case of the partial light absorption and high solute conversion, variation (decrease) of the solution absorbance - at the irradiation wavelength - with the irradiation time has to be established.

Provided that the product of the photochemical reaction does not absorb light at the irradiation wavelength, or its absorption is negligible in comparison with the absorption by the actinometric solute, the following equations for calculating incident photon flow Φ_p can be derived [31]:

$A_\lambda \geq 2$

$$\Phi_p = \frac{\Delta n_{Ac}}{\Phi \cdot t}$$

$A_\lambda < 2$, low conversion

$$\Phi_p = \frac{\Delta n_{Ac}}{\Phi \cdot t \cdot (1 - 10^{-\overline{A_\lambda}})}$$

$A_\lambda < 2$, high conversion

$$\Phi_p = \frac{N \cdot V}{2.303 \cdot l \cdot \Phi \cdot \varepsilon_\lambda} \cdot a$$

a = slope of the dependence $\ln(10^{A_{\lambda t}} - 1)$ vs. irradiation time t

Φ_p = incident photon flow [s^{-1}]
Δn_{Ac} = number of molecules of solute consumed or product formed
Φ = quantum yield
t = irradiation time [s]
$\overline{A_\lambda}$ = time averaged absorbance of the actinometer solution in irradiation cell during irradiation time t
ε_λ = molar absorption coefficient of the actinometer solute at the wavelength of irradiation λ [dm^3 mol^{-1} cm^{-1}]
l = optical path length of the irradiation cell [cm]
V = volume of the irradiated solution [dm^3]
$A_{\lambda t}$ = absorbance of the actinometer solution in the irradiation cell at a time t
N = Avogadro number

When the light absorption by the photochemical conversion product is not negligible, photokinetic equations lead to more complex expressions, which have to be solved numerically [33-35].

In order to prevent the "inner filter effect", i.e. the absorption of light by the product over a short distance in the actinometer solution layer facing the irradiation source, good stirring of the solution should be provided, and the concentration of the actinometer should not be too high, especially when the measurement is carried out with full light absorption.

In principle, any photochemical reaction for which quantum yield is reliably known can serve as an actinometer. However, when choosing an appropriate system, there are some other criteria, which should be considered [31-33]: quantum yield should be relatively insensitive to changes in wavelength, temperature, reactant concentration, presence of oxygen and trace impurities, and changes in radiation intensity; also, chemical analysis of photochemical conversion should be relatively simple, when a spectrophotometric analysis is preferred. Some actinometric solutions eligible for UV light measurement are summarized in Table VIII.

Table VIII Chemical Actinometers for UV Light Measurement

Actinometer, Solute concentration	Wavelength Range [nm]	Quantum yield Φ	Photoconversion analysis ε_λ [cm^{-1} dm^3 mol^{-1}][a]	Ref.
Potassium ferrioxalate 0.006-0.15 mol dm^{-3} in 0.05 mol dm^{-3} H$_2$SO$_4$	254-400	1.1-1.25	Fe II determination via complex with phenanthroline $\varepsilon_{510} = 1.11 \cdot 10^4$	(32,36)
Uranyl oxalate 1 (or 10) mmol dm^{-3} UO$_2$SO$_4$ 5 (or 50) mmol dm^{-3} H$_2$C$_2$O$_4$	200-400	0.5-0.6	Oxalate loss by titration or by spectrophotometry	(32,36)
Uridine $10^{-4} - 10^{-3}$ mol dm^{-3}	≈ 220-280	≈ 0.019	Chromophore loss $\varepsilon_{262} = 1.04 \cdot 10^4$	(37,38)
3,4-Dimethoxynitro-benzene (DMNB) 10^{-4} mol dm^{-3} in 0.5 mol dm^{-3} KOH	254-365	0.116	Determination of the photohydrolytic product at 450 nm $\varepsilon_{450} = 3 \cdot 10^3$	(39)
Malachite green leucocyanide (MG-CN) 10^{-4}–10^{-3} mol dm^{-3} and 10^{-3} mol dm^{-3} HCl in ethanol	225-330	0.9-1.0	Determination of the coloured form (MG$^+$) at 620 nm $\varepsilon_{620} = 1.06 \cdot 10^5$	(32,40)

[a] molar absorption coefficient at wavelength λ (nm) of the spectrophotometric analysis

The UV optical absorption spectra of the solutions are shown in Figure 6; (ferrioxalate solution at a concentration of 0.02 mol dm^{-3} with optical pathlength 1 cm shows 100 % absorption at wavelengths shorter than 400nm).

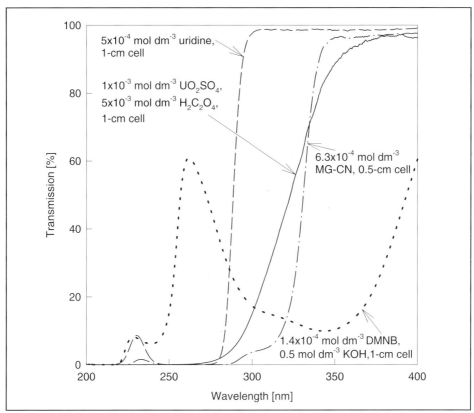

Figure 6 Optical Absorption Spectra of Actinometric Solutions

The potassium ferrioxalate actinometer, based on the photoreduction of ferric ions to ferrous ions, is apparently most frequently used. As it absorbs light also in the visible range, it has to be handled in the dark. In the wavelength range below 250 nm, considerably lower quantum yield values have been reported (0.68, 0.67 and 0.50 at 240, 230 and 222 nm, respectively) [41]. Similarly, the quantum yield also seems to be decreased on irradiation with intense 193-nm UV light from an ArF excimer laser, and in this case the presence of oxygen in solution also plays a significant role, apparently due to the reaction of ferrous ions with ozone [42]. For these reasons, the use of ferrioxalate actinometers below 250 nm cannot be recommended unless quantum yield values are reliably established.

In uranyl oxalate actinometer, the UO_2^{2+} ion plays the role of a sensitizer, while unsensitized photolysis of oxalic acid proceeds with a considerably lower quantum yield. When using this actinometer at short UV wavelengths (e.g. 222 nm), the use of lower solute concentrations (0.001 mol dm^{-3} UO_2SO_4 and 0.005 mol dm^{-3} $H_2C_2O_4$) and efficient solution mixing is recommended, in order to prevent the inner filter effect - due both to oxalic acid and uranyl sulphate (in the case of oxalate depletion) [43].

If a uridine solution is used for 222-nm light actinometry, the quantum yield is apparently close to 0.019, but light absorption by the photohydrate formed has to be taken into account. The latter was neglected in a recent application of this actinometer to a 222-nm excimer lamp, when a high and unjustified value of quantum yield of 0.034 was used [44]. However, considerably higher quantum yields of the chromophore loss (depending on the presence or absence of oxygen) were indeed observed on 193-nm laser irradiation, due to the processes initiated by photoionization [45].

Actinometry with an alkaline 3,4-dimethoxynitrobenzene solution is conveniently carried out at partial light absorption. In this case, corrections have to be made for UV-light absorption by the coloured photohydrolytic product, 2-methoxy-5-nitrophenolate anion [39].

A dyed cation formed on irradiation in malachite green leucocyanide solutions also absorbs UV light (the absorption maximum is at $\approx$ 317 nm, and the molar absorption coefficient between 290 and 340 nm is higher than that of the leucocyanide). Therefore the possibility of the inner filtration has to be considered at high irradiance levels.

Actinometry in polychromatic light should be carried out with a system which has a quantum yield independent of the wavelength, and a solute concentration should be chosen so that there is total light absorption over the spectral domain being considered. Unless these conditions are fulfilled, the relative spectral distribution is also needed for calculating the photon flow. For integrating light from visible (750 nm) down to about 313 nm, dilute acid solutions of Reinecke's salt $KCr(NH_3)_2(SCN)_4$ have been suggested [31,32]. SCN^- ions are formed with a quantum yield $\Phi \approx 0.3$, and their concentration is determined spectrophotometrically, via a coloured complex with ferric ions. Possibly also potassium ferrioxalate and uranyl oxalate actinometers might serve the purpose in the spectral ranges of 250-400 and 200-300 nm, respectively.

4. Comparison of radiometer with chemical actinometer

Radiometers are certainly the only devices which can be used for routine measurements with the powerful UV radiation sources for curing, however there are two problems which may be connected with their use. Firstly, photodiodes can suffer changes in sensitivity (generally decrease) after high radiant power exposures, and that is why occasional recalibration against a standard is strongly recommended. The second problem is connected with radiometer calibration itself. Due to the spectral dependencies of the semiconductor diodes which are currently used in radiometers, serious discrepancies have been observed and reported when comparing radiometers with a spectro-radiometer, if the spectrum of the measured radiation was different from that used for radiometer calibration [46,47]. Therefore it is recommended that the meters are calibrated for the source they have to measure, and that recalibration is performed after a modification of the spectral distribution of a source.

A certain solution to the problem seems to be offered by actinometry, which might be used to check or calibrate the radiometer. Actinometric measurements usually have to be carried out under restricted power of the light source, and light intensity can be further reduced by increasing the distance from the source or by using a rotating segment. The actinometric solution can simply be irradiated from the top in an open Petri dish, under efficient mixing with an electromagnetic stirrer.

Thus the potassium ferrioxalate and uranyl oxalate actinometers were used for checking the URADES 4 (from G.E.R.U.S. mbH, Berlin) readings in the light of the 308-nm and 222-nm excimer lamps (Heraeus Noblelight), respectively. In both cases the light was fully absorbed in the actinometric solution. When irradiating the actinometer with the 222-nm lamp, the whole solution surface was irradiated. In the case of the 308-nm lamp only a small fraction of the solution surface was irradiated through a simple collimator, so that the exposure times could be extended.

Relating the photon flow measured by actinometry to the irradiated solution surface, photon exposure values were obtained; then taking the effective photon energy as corresponding to 222 nm or 308 nm wavelengths, radiant exposure values were calculated. These are plotted in Figure 7 against radiant exposure values, as calculated from the irradiance measured under the same irradiation conditions with URADES 4 for individual exposure times. The agreement between the results is obvious.

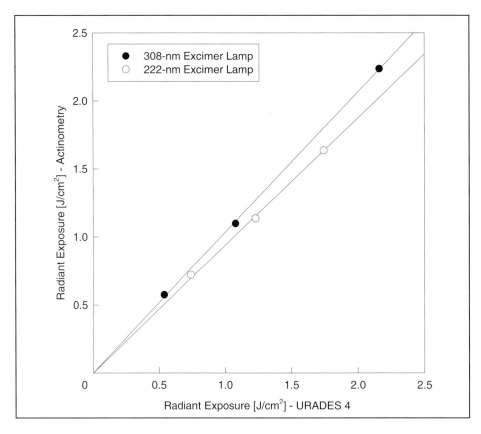

Figure 7 Comparison of the Radiant Exposure Values as Obtained from Irradiance Measured with URADES 4 with Chemical Actinometry

5. UV-sensitive films and labels

UV-sensitive films and labels, when properly calibrated, can be used for monitoring the dose which the material receives during the curing process, and they are particularly practical in inaccessible curing systems. They can be conveyed on the belt or attached to the cured material and passed through the web system - web offset, flexo, screen and other systems - at the proper cure rate. The dose readings can be used for evaluating curing system performance and also lamp degradation.

UV Process Supply, Inc. markets a CON-TROL-CURE Rad Check UV Measurement System for measuring UV dose in the spectral range of 320-380 nm. It consists of two types of Test Strips, 300 and 800, for measuring UV doses up to 300 mJ/cm^2 and 800 mJ/cm^2, respectively, and of calibrated read-out instruments, Rad Check Dosimeters 300 and 800. The test strips contain a UV/EB sensitive compound (on a polyester support) which is destroyed both by UV-light and EB-irradiation.

The same company supplies UV/EB Intensity Labels which, when exposed to UV light or ionizing radiation, change colour gradually from green to purple. Their use requires the in-plant creation of a calibration standard consisting of labels irradiated in incremental steps to different dose levels.

Radiochromic films containing different types of leucocyanides of triphenylmethane dyes (e.g. pararosaniline cyanide, hexahydroxyethyl pararosaniline cyanide, new fuchsin cyanide) show colouration after UV exposure with wavelengths in the range from 200 to ≈ 350-400 nm, film sensitivity being spectral dependent [48,49]. The sensitivity of the Risø B3 radiochromic film irradiated with 222-nm or 308-nm excimer lamps appears to be quite appropriate for measuring UV doses in the range of ≈ 0.05-1.5 J/cm^2. The film response is not significantly intensity dependent, however, it is strongly influenced by fractionation of the UV dose. When the dose is delivered in separate fractions, a considerably smaller response is observed than when the dose is delivered all at once; this might be due to some relatively slow post-irradiation reactions which change the film UV-absorption properties, causing inner filtration. Radiochromic films are likely to be applied to the "cold lamps" only.

Some UV-sensitive films which have been developed primarily for biological and phototherapy purposes [47] could also be used in some industrial UV-light applications. Thus the commercially available film containing a diazonium compound is sensitive to UVA (315-400 nm) [50]. After exposure, the film is developed in ammonia vapour where it turns blue due to the colour reaction of the undecomposed diazonium compound. For measuring UVB radiation, thermoplastic, 40-μm thick polysulphone film has been used [51]. After UV irradiation an increase in film optical absorbance at 330 or 300 nm is measured. None of the films has probably been tested for the effects of high temperatures, which might limit their applicability.

IV. REFERENCES

1. W. L. McLaughlin, A. W. Boyd, K. H. Chadwick, J. C. McDonald, A. Miller: Dosimetry for Radiation Processing, Taylor & Francis, London, 1989.
2. ASTM Standard E 1818-96: Standard Practice for Dosimetry in an Electron Beam Facility for Radiation Processing at Energies Between 80 and 300 keV, American Society for Testing and Materials, Philadelphia, 1996.
3. F. H. Attix: Introduction to Radiological Physics and Radiation Dosimetry, John Wiley & Sons, New York, 1986.
4. J. R. Greening: Fundamentals of Radiation Dosimetry, Adam Hilger Ltd., Bristol, 1981.
5. ICRU Report 33 Radiation Quantities and Units, International Commission on Radiation Units and Measurements, Bethesda, Maryland, 1980.
6. ICRU Report 35 Radiation Dosimetry: Electron Beams with Energies Between 1 and 50 MeV, International Commission on Radiation Units and Measurements, Bethesda, Maryland, 1984.
7. S. M. Seltzer, M. J. Berger: Int. J. Appl. Radiat. Isot. 33, 1189 (1982).
8. S. M. Seltzer, M. J. Berger: Int. J. Appl. Radiat. Isot. 35, 665 (1984).
9. W. L. McLaughlin, J. C. Humphreys, B. B. Radak, A. Miller, T. A. Olejnik: Radiat. Phys. Chem. 14, 535 (1979).
10. W. L. McLaughlin, R. M. Uribe, A. Miller: Radiat. Phys. Chem. 22, 333 (1983).
11. W. L. McLaughlin, J. C. Humphreys, D. Hocken, W. J. Chappas: Radiat. Phys. Chem. 31, 505 (1988).
12. A. Miller, W. Batsberg, W. Karman: Radiat. Phys. Chem. 31, 491 (1988).
13. M. C. Saylor, T. T. Tamargo, W. L. McLaughlin, H. M. Khan, D. F. Lewis, R. D. Schenfele: Radiat. Phys. Chem. 31, 529 (1988).
14. R. D. H. Chu, G. Van Dyk, D. F. Lewis, K. P. J. O'Hara, B. W. Buckland, F. Dinelle, Radiat. Phys. Chem. 35, 767 (1989).
15. W. L. McLaughlin, Chen Yun-Dong, C. G. Soares, A. Miller, G. Van Dyk, D. F. Lewis: Nucl. Instr. Methods A302, 165 (1991).
16. ASTM Standard E 1650-94: Standard Practice for Use of a Cellulose Acetate Dosimetry System, American Society for Testing and Materials, Philadelphia, 1994.
17. R.. Tanaka, S. Mitomo, H. Sunaga, K. Matsuda, N. Tamura: JAERI-M Report 82-033, Japan Atomic Energy Research Institute, Tokyo, 1982.
18. W. L. McLaughlin, B. Wei-Zhen, W. J. Chappas: Radiat. Phys. Chem. 31, 481 (1988).
19. I. Janovsky, K. Mehta: Radiat. Phys. Chem. 43, 407 (1994).
20. W. L. McLaughlin, J. M. Puhl, A. Miller: Radiat. Phys. Chem. 46, 1227 (1995).
21. A. A. Abdel-Fattah, A. Miller: Radiat. Phys. Chem. 47, 611 (1996).
22. D. F. Regulla, U. Deffner, Int. J. Appl. Radiat. Isotopes 33, 1101 (1982).
23. ASTM Standard E 1607-94: Standard Practice for Use of the Alanine-EPR Dosimetry System, American Society for Testing and Materials, Philadelphia, 1994.

24. ASTM Standard E 1631-94: Standard Practice for Use of Calorimetric Dosimetry Systems for Electron Beam Dose Measurements and Dosimeter Calibrations, American Society for Testing and Materials, Philadelphia, 1994.
25. H. Sunaga, Y. Haruyama, H. Takizawa, T. Kojima, K. Yotsumoto, Proc. of the 6th Japan-China Bilatelar Symposium on Radiation Chemistry, Waseda University, Tokyo, Japan, Nov. 1994, JAERI-Conf 95-003.
26. I. Janovsky, A. Miller: Appl. Radiat. Isot. 38, 931 (1987).
27. W. L. McLaughlin: Dosimetry for Low-Energy Electron Machine Performance and Process Control, Proceedings of RadTech '90 North America, Chicago, March 1990, Vol. 2, p.91, Radtech International North America, Northbrook, Il, 1990.
28. T. Tabata, K. Shinoda, P. Andreo, Wang Chuanshan, R. Ito: RadTech Asia '93 Conf. Proc., Nov. 1993, Tokyo, Japan, p. 574, RadTech Japan Organizing Committee, Tokyo (1993).
29. T. Tabata: Bull. Univ. Osaka Prefecture A 44, 41 (1995).
30. Commission on Photochemistry, Organic Chemistry Division, International Union of Pure and Applied Chemistry, Pure & Appl. Chem. 68, 2223 (1996).
31. A. M. Braun, M-T. Maurette, E. Oliveros: Photochemical Technology, p.51, John Wiley & Sons, Chichester, England, 1991.
32. Commission on Photochemistry, Organic Chemistry Division, International Union of Pure and Applied Chemistry, Pure & Appl. Chem. 61, 187 (1989).
33. N. J. Bunce: Actinometry, in Handbook of Organic Photochemistry, Vol. I, p. 241, J. C. Saciano ed., CRC.Press, Inc., Boca Raton, Florida, 1989.
34. G. Gauglitz, S. Hubig: Z. Phys. Chem. N. F. 139, 237 (1984).
35. G. Gauglitz: Stud. Org. Chem. (Amsterdam) 40 (Photochromism, Mol. Syst.), 883 (1990).
36. S. L. Murov, I. Carmichael, G. L. Hug: Handbook of Photochemistry, 2nd ed., Marcel Dekker, Inc., New York, 1993.
37. C. von Sonntag, H-P. Schuchmann: J. Water Sci. Res. Technol. – Aqua 41, 67 (1992).
38. G. J. Fisher, H. E. Johns: Pyrimidine Photohydrates, in Photochemistry and Photobiology of Nucleic Acids, Vol. 1, p.169, ed. S. Y. Wang, Academic Press, New York, 1976.
39. L. Pavlickova, P. Kuzmic, M. Soucek: Collection Czechoslovak Chem. Commun. 51, 368 (1985).
40. J. G. Calvert, H. J. L. Rechen: J. Am. Chem. Soc. 74, 2101 (1952).
41. E. Fernandez, J. M. Figuera, A. Tobar: J. Photochem. 11, 69 (1979).
42. Y. Izumi, K. Ema, T. Yamamoto, S. Kawanishi, Y. Shimizu, S. Sugimoto, N. Suzuki: The Review of Laser Engineering 19, 247 (1991).
43. G. S. Forbes, L. J. Heidt: J. Am. Chem. Soc. 56, 2363 (1934).
44. J. –Y. Zhang, I. W. Boyd, H. Esrom: Appl. Surf. Sci. 109/110, 482 (1997).
45. G. G. Gurzadyan, H. Görner: Photochem. Photobiol. 60, 323 (1994).
46. R. M. Sayre, L. H. Kligman: Photochem. Photobiol. 55, 141 (1992).

47. B. L. Diffey: *New Trends in Photodosimetry, Cosmetic Science and Technology Series, Vol. 10, Sunscreens: Development, Evaluation, and Regulatory Aspects,* p. 93, ed. N. J. Lowe and N. A. Shaath, Marcel Dekker, Inc., New York, 1990.
48. F. Abdel-Rehim, S. Ebrahim, A. A. Abdel-Fattah, J. Photochem. Photobiol. A Chem., 73, 247 (1993).
49. F. Abdel-Rehim, A. A. Basfar, A. Abdel-Fattah, J. Photochem. Photobiol. A Chem. 101, 63 (1996).
50. S. A. Jackson: J. Biomed. Eng. 2, 63 (1980).
51. B. L. Diffey: *Ultraviolet Radiation Dosimetry with Polysulphone Film, Radiation Measurements in Photobiology,* p. 135, ed. B. L. Diffey, Academic Press, London, 1989.

CHAPTER VI

ELECTRON BEAM (EB) CURING EQUIPMENT

I. GENERATION OF ACCELERATED ELECTRONS

1. Principle of electron acceleration

Electrons which are able electronically to excite and to ionise organic molecules such as acrylates or epoxides must have energies in the range of 5-10 eV. Such electrons can be produced from fast electrons as a result of the energy degradation process in liquids, solids or gasses. The secondary electrons generated from the primary electron show an energy distribution with a maximum in the range between 50 and 100 eV. In contrast to the fast electrons exhibiting energies in the keV to MeV range, the penetration depth of secondary electrons in liquids or solids only reaches a few nanometers. Thus, secondary electrons generate ions, radicals and excited molecules in "droplets" along the track of the fast electron. Finally, electrons with energies below the threshold of molecular vibrational excitation lose their energy by rotational excitation and finally become thermalised. Samuel and Magee [1] denoted such droplets containing several pairs of ions, radicals and excited molecules as "spurs" (see Figure 1).

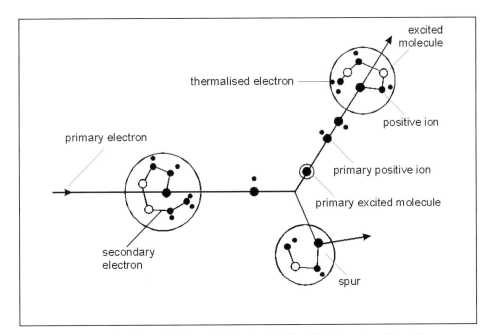

Figure 1 "Spurs" Along the Path of a Fast Electron in a Liquid

Radiation chemical processing of monomers and polymers such as electron beam curing, cross-linking, grafting and degradation of polymers is induced by different chemically reactive species formed initially by electrons in the spurs [2,3,4].

In curing applications, electrons have to penetrate reactive liquids with typical masses per unit area of one to several hundred gm^{-2} (1 gm^{-2} = 1 µm at unit density). As shown in the electron depth dose profiles given in Chapter I, electrons with energies >100 keV must be used for curing. On a technical scale, such electrons are generated by means of single gap, low-energy electron accelerators.

The principle of fast electron generation is very simple (see Figure 2): electrons are generated in a vacuum by a heated cathode. The electrons emitted from the cathode are accelerated in the electrostatic field applied between cathode and anode. Acceleration takes place from the cathode, which is on negative high voltage potential to the grounded accelerator vessel as anode. Often an electron optical system is used to focus the accelerated electrons to the accelerator window plane.

The energy gain of the electrons is proportional to accelerating high voltage. It is given in electron volts (eV), i.e. the energy which gaining a particle of unit charge by passing a potential difference of 1 V. The electrons leave the vacuum chamber and reach the process zone if their energy is high enough to penetrate the 15-20 µm thick titanium window foil.

Additionally, stopping of electrons in matter leads to X-ray generation. Therefore, the electron accelerator and process zones have to be shielded to protect the operator. For electron energies up 300 keV this can be done by self-shielding these units using lead cladding.

VI – Electron Beam (EB) Curing Equipment   137

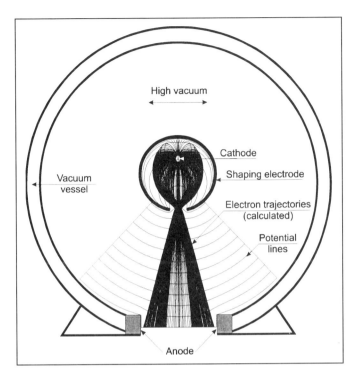

Figure 2   Principle of Electron Acceleration in a Single Gap Low-Energy Accelerator

Based on the principle illustrated in Figure 2, there are three fundamental designs used for industrial low-energy electron accelerators: point electron sources with scanned beam [5], single or multiple linear emitters placed perpendicularly to product direction [6,7,8,9,10,11], and a multiple emitter assembly with filaments placed in parallel to product direction [12,13,14] (see Figure 3).

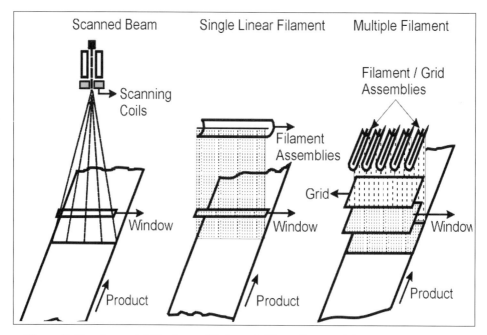

Figure 3 Fundamental Designs of Industrial Low-Energy Electron Accelerators

The low-energy electron accelerators based on these designs are now well developed and can be used as reliable computer-controlled subsystems in coating lines, printing presses, laminating machines etc. Their operating parameters such as electron energy, electron beam power, irradiation width and delivered dose rate can be matched to the demands of the industrial curing process. For economic and technical reasons, process "dedicated" electron accelerators are applied. The term "electron processor" is often used for such machines.

The selection of a suitable electron processor type is dictated by process parameters. Scanning type electron accelerators use a typical triode electron gun which generates and shapes a pencil beam. Using periodically changing magnetic fields the beam is deflected over the window area. Carefully tuned deflection leads to a very uniform current (dose rate) distribution across the beam exit. Energies up to 300 keV can be reached at medium to high beam current.

Due to the limited deflection angle, the scanning horn must have a length which is somewhat larger than the width of the exit window.

These dimensional problems are avoided by other types of electron accelerators using cylindrical acceleration chambers which can be kept very compact. In a second accelerator design, beam current and width are determined by a linear filament. Since the current extractable from a unit length of the heated filament is limited, multi-filament configurations are often applied. This also improves the uniformity of the current distribution, which is usually less homogeneous than for scanning type machines. Beam width is limited by the necessity to span the filament across the full exit width. However, a beam width of about 2 m can be obtained.

Many problems related to filament support and adjustment can be avoided if an assembly of short (about 20 cm) filaments is set parallel to the product direction. The filaments are mounted between two rigid bars determining the length of the filament assembly. For each filament, a control and a screen grid define the electron optical extraction conditions into a single accelerator gap. Overlapping of the beams emitted from adjacent cathodes leads to the formation of a uniform electron "cloud" before acceleration takes place. Great beam width, high beam power and good dose uniformity are obtained on the basis of this design principle.

2. Characteristics of EB curing in comparison with thermal and UV curing

Progress of EB curing in recent years, as well as its future development, are determined by its ability to compete successfully with conventional curing processes and other radiation techniques such as UV curing. Usually, the most important reason for a company to introduce EB curing is not technology based. It is the possibility of creating unique or vastly improved products at lower cost. Electron beam curing should not exclusively be regarded as a processing technology, where the electron processor is the main element. Although EB shows striking advantages over conventional curing process technologies, such as reduced energy consumption, lower space requirements, considerably lower substrate heating, higher curing speed, improved control and pollution-free operation when using solvent-free coatings, the final properties of the resulting product are essentially determined by the formulation of the coating, the properties of the substrate and the characteristics and functionality of the cured coating. Many of the above-mentioned advantages of EB curing also apply to UV curing. A brief analysis of advantages and disadvantages of UV and EB curing was already given in Chapter I, Table V.

The development of EB and UV curing in the past 40 years suggests that they are complementary technologies. However, when it comes to variety of application and production capacity, UV has been shown to have immediate and increasingly valid advantages within its own specific areas of application. Its growth therefore is strong but so is that for Electron Beam in its own specialised areas of application.

For many solvent or water-based coatings, there is still no solvent-free alternative showing comparable or better functionality. Additionally, retrofitting of existing conventional curing lines with EB (or UV) equipment is not possible in most cases, as specific application equipment is required for solvent-free systems. Thus the future success of radiation curing will require dedicated efforts from all parties contributing to the technology: raw material suppliers, formulators, machine manufacturers, end users as well as research institutes

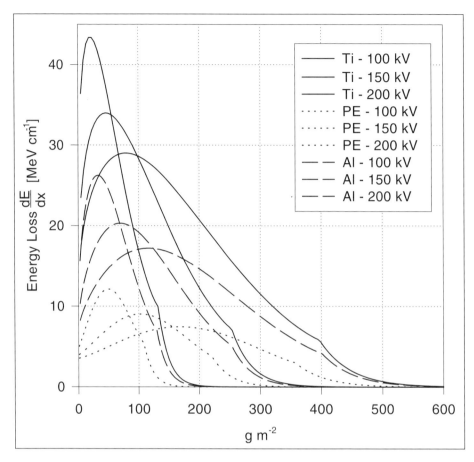

Figure 4 Depth Dose Distribution and Different Absorbers [16]

The electron penetration range is related to the path length which the electron travels during the energy degradation process. It can be estimated from depth dose distribution. Figure 5 gives the electron penetration range (in gcm^{-2}) as a function of electron energy. This relationship applies for absorbers with low atomic numbers and atomic weights, e.g. polyethylene, and has to be modified for heavier absorbers.

3. Definitions and units

Before looking at the technical design and characteristics of various industrial electron processor units, it might be useful to give some unit definitions and explain the terminology of electron beam processing. The basic electrical parameters of an electron processor are its acceleration voltage, the electron beam current and the electron beam power. The ratio of electron beam power to the input electrical power defines the efficiency of an electron accelerator. The acceleration voltage determines the energy of the electrons.

A second group of definitions is related to the absorption properties of the accelerated electrons in matter: the absorbed dose, the depth dose profile, the penetration range and the dose rate.

In a third group of terms, process parameters such as the line speed are included. If dose rate and line speed are combined, the dose delivered to the product to be cured can be calculated. A relationship between beam current and line speed is obtained by introducing a processor specific yield factor. Often the dose-speed capacity of a processor is given by the manufacturer. This is the product of the line speed times the delivered dose at maximum electron beam power.

A brief definition of the units mentioned above is given below:

Acceleration voltage: Potential difference between cathode and anode of the accelerator usually measured in kV at a voltage divider chain of the high voltage unit.

Electron beam current: Number of electrons per second emitted from the cathode, measured in mA (1 mA = 6.25×10^{15} electrons per second) at the high voltage unit.

Electron beam power: Product of the acceleration voltage times the electron beam current, measured in kW (1 kW = 10 mA × 100 kV).

Absorbed dose: Mean energy imparted by ionising radiation to matter in a volume element divided by the mass of that volume element (definition from ICRU [15]). Thus, the absorbed dose is expressed as unit energy per unit mass: 1 Gray (Gy) = 1 Jkg^{-1}. (Until 1986 the dose unit Megarad (Mrad) was used: 1 Mrad = 10 kGy). The dose can be measured by using calorimeters. However, calibrated foil dosimeters are frequently applied for low-energy electron processors (see Chapter V).

Depth dose profile: The energy deposition produced by electrons of a given energy in an absorber can be expressed as a function of the depth, the atomic number and the atomic weight of the absorber. Depth dose profiles calculated for electron energies between 100 and 200 keV [16] are shown in Chapter I. The energy loss dE/dx of the fast electrons (expressed in MeV cm^{-1}) is given as a function of the mass per unit area (gm^{-2}). It is directly proportional to the absorbed dose: $1/\rho dE/dx$ = dose × unit area, or in dimensions $Jg^{-1}cm^2$.

Figure 4 shows an example of depth dose profiles, calculated for different absorbers. In a realistic electron irradiation process, energy losses occur in the titanium window foil, the air gap between foil and coating, within the coating and in the substrate (see Figure 5 of Chapter V).

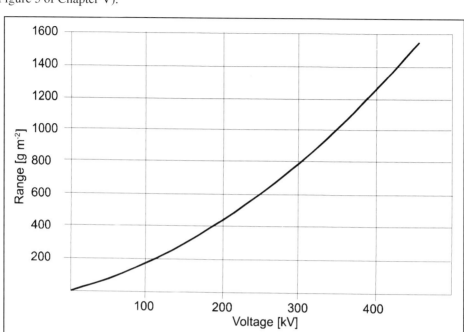

Figure 5 Electron Penetration Range Versus Electron Energy

Dose rate: Dose per unit time, measured in $Gys^{-1} = J\ kg^{-1}s^{-1}$. At a constant acceleration voltage it is proportional to the electron beam current:

$$\text{Dose Rate} \sim \text{Beam Current}.$$

Line speed: Speed of the product to be irradiated, measured in m/min. It determines the exposure time of the product.

Delivered dose: Ratio of dose rate divided by the line speed:

$$\text{Delivered Dose} = \text{Dose Rate/Exposure Time}$$

At a given dose rate the line speed is often adjusted to obtain the desired dose. On the other hand, the beam current can be controlled by the line speed to maintain a constant delivered dose, especially during the start up of curing.

Yield factor: This term is used to characterise the curing performance of an electron processor. It is a constant that relates the delivered dose to the beam current and line speed:

$$\text{Delivered Dose} = \text{Yield Factor} \times \text{Beam Current/Line Speed}.$$

The yield factor is measured in $kGy \times m/(min\ mA)$.

Dose-speed capacity: Product of the delivered dose times the line speed at maximum beam power:

Dose-Speed Capacity = Delivered Dose × Line Speed.

The dose-speed capacity is practically measured in m/min at 10 kGy. It is the most convenient unit to relate the curing performance of an electron processor to the desired process parameters dose and line speed. Figure 6 illustrates this relationship by way of example. At a given delivered (surface) dose and acceleration voltage, the maximum line speed can be taken from a nomogram often provided by the processor manufacturer.

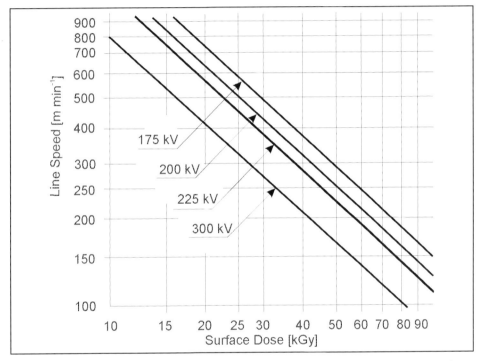

Figure 6 Line Speed as a Function of the Delivered Dose at Different Acceleration Voltages

II. TYPES OF INDUSTRIAL LOW-ENERGY ELECTRON PROCESSORS

1. Accessible electron energy and beam power range

For single gap electron accelerators, the electron energy is limited to about 300 keV. Even if high voltage insulation problems could be avoided, e.g. by increasing the distance between cathode and anode, resulting nonlinearities of the accelerating potential would prevent a stable accelerator operation. Therefore, accelerator tubes with many electrodes are used at higher voltages. This leads to completely different, more expensive accelerator types and renders radiation self-shielding by lead cladding very difficult or impossible.

There are also physical limitations for the maximum beam power:

Fast electrons penetrating an absorber, e.g. the titanium window foil of an electron accelerator, lose part of their energy by collisions and radiation. The electron energy spectrum is changed: the maximum shifts to lower energy and energy distribution is broadened. Energy shift and broadening are less important for the curing process. However, energy deposition within the window foil determines the maximum achievable beam power of a low-energy electron accelerator.

Energy absorption in the foil can be calculated from the depth dose distribution generated by electrons traversing the foil. Figure 7 gives energy absorption as a function of the electron energy for a 10 µm thick titanium foil [17].

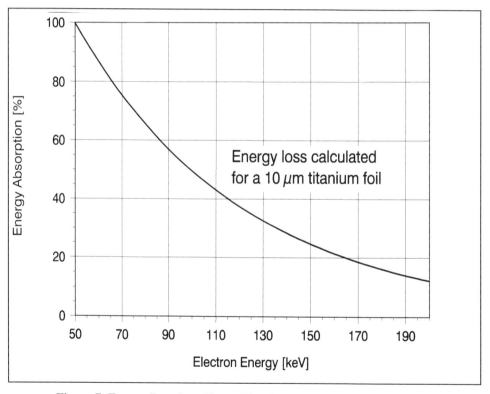

Figure 7 Energy Loss in a 10 µm Titanium Foil as a Function of the Electron Energy

For a 10 μm titanium foil, energy absorption increases from 12% at 200 keV to 25% at 150 keV and reaches 50% at 100 keV. This energy loss and the heat conductivity properties of the window support determine the foil temperature. The foil temperature is the most critical parameter limiting the beam power density (expressed in W cm^{-2}) of a low-energy electron accelerator.

Solving a three dimensional heat conduction equation, temperature profiles were calculated for a 15 μm titanium foil covering a rectangular mesh unit of the copper window support flange. Assuming a 15 μm titanium foil, a beam power density of 40 W cm^{-2} and a window transmission of 70%, temperature profiles were calculated for the mesh unit geometry indicated in Figure 8.

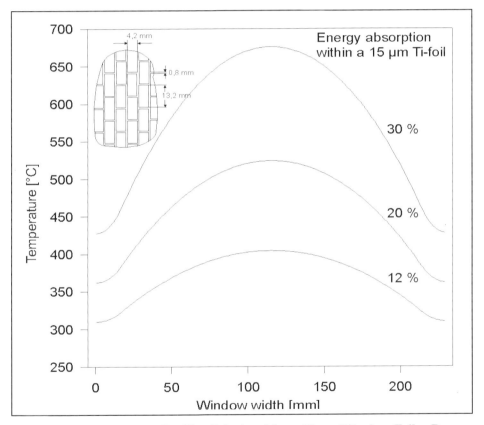

Figure 8 Temperature Profiles Calculated for a 15 μm Titanium Foil, a Power Density of 40 W cm^{-2} and 70% Window Transmission. Mesh Structure of the Copper Support as Indicated

Figure 8 gives the energy absorption within the window foil as a parameter. Window foil temperatures are difficult to measure. However, the colouration of the foil can be used as temperature indicator in the interesting temperature range. A weak brownish colour appears at about 400°C which turns into metallic blue at 500°C. As maximum foil temperatures of more than 400 - 450°C cannot be tolerated (see lowest temperature profile in Figure 8), it is obvious that the power density must be reduced by at least a factor of two when 30% of the electron energy is absorbed in the window foil. This is the energy loss calculated for 150 keV electrons penetrating a 15 µm titanium foil. As a rule, the tolerable maximum current density on the foil decreases from 0.2 mA cm^{-2} at 200 keV to 0.1 mA cm^{-2} for 150 keV electrons and to 0.07 mA cm^{-2} at 120 keV.

Thus, the design of the exit window is of crucial importance for accelerator performance, in particular the maximum beam current. This can be illustrated by the following example:

Parallel to the product direction, the width of the exit window is assumed to be 20 cm. At 200 keV, this corresponds to a maximum beam current of 4 mA per cm of window length, which decreases to 2 mA /cm at 150 keV and 1.4 mA/cm at 120 keV.

The limited current density has severe consequences for possible applications of low-energy electron accelerators.

There are presently three major application fields imposing different demands on electron accelerators for curing. Electron beam curing of coatings on rigid substrates, mainly doors, parquet, laminated panels, roof tiles, plastic and metal sheets, requires the full range of electron energies. Some of these coatings are on non-porous substrates and reach only masses per unit area of less than 50 g m^{-2}. Electron energies in the range of 150 keV are sufficient for these applications. On the other hand, curing of higher coating masses on porous substrates and curing of three-dimensional parts need electron energies up to 250 keV or even more. In the case of rigid substrates, production speed is not determined by the curing step but rather by other operations such as feeding and substrate pretreatment. Therefore, only low or moderate electron beam powers are required. This is illustrated in the upper part of Figure 9.

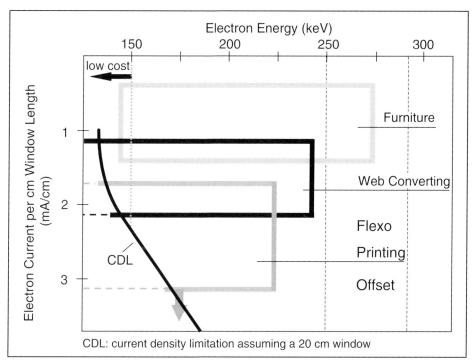

Figure 9 Curing Application Fields of Low-Energy Electron Accelerators, CDL = Current Density Limitation Assuming a 20 cm Window

Electron beam curing of coatings on flexible substrates takes place from roll to roll. Production speeds of 100 - 300 m/min are common, with coating weights from 1 to 30 g m^{-2}. This requires low energies at moderate to high beam powers. As shown in the middle part of Figure 8, an electron energy of 150 keV would be enough for most applications. However, the current density limitation could be approached with this energy. This can be avoided by increasing the electron energy. The dotted line shows energies which are required when electrons have to penetrate foils or paper to reach the coating. Laminating and cast coating are such examples.

In web offset printing presses reach machine speeds of up to 1000 m/min. Although the real printing speeds are usually much lower, very high beam powers are required. The electron energy could be kept very low. In practice, the lower limit of the electron energy is set by the current density limitation.

2. Low-energy electron accelerators used in industry

First attempts to use electrons for curing of coatings date back to the late 1950s. However, the lack of adequate chemistry prevented rapid development of electron beam curing. Polyester/styrene systems, at that time the only systems available on a technical scale, had to be prepared at processing temperatures of up to 200°C and allowed only low cure speeds. In addition, UV curing showed better results than electron beam curing. The result was that the latter was restricted to wood coatings.

With the development of epoxy and later urethane acrylates in the 1970s, the original objective of radiation curing, i.e. to provide in situ polymerisation and cross-linking at low cost, could be approached. The subsequent introduction of reactive diluents such as acrylate monomers, enabled the formulation of a large variety of different radiation curable coatings, inks and varnishes. This development stimulated the design of a new generation of low-energy, self-shielded electron acceleratores for radiation processing: electron processors for curing applications. Energy Science Inc. (ESI) invented an accelerator using a linear cathode and a single acceleration gap within a cylindrical vacuum chamber [18]. In comparison with bulky earlier units, the dimensions of the new electron accelerator, named Electrocurtain ®, were reduced substantially. At the same time, Polymer Physik [19] developed a single gap, vacuum insulated, self-shielded electron scanner with small dimensions. Later RPC Industries [20] introduced a multi-filament linear cathode electron BroadBeam® processor, thus opening up the field of high speed curing applications. The Japanese curing market was mainly served by Nissin High Voltage (NHV) manufacturing scanning as well as linear cathode type electron processors. Despite the relatively high investment costs of EB machines and the complicated technologies involved such as high voltage, high vacuum and radiation protection, a steady progress in electron beam curing applications was obtained. At the end of the 1980s, more than 200 pilot and production electron processors were in operation worldwide. Whilst electron beam curing has many theoretical advantages compared with thermal drying or UV curing, the commercial expectations related to this technology did not come true. Electron processor manufacturers analysed the commercial situation in the 1990s and responded in two different ways: increasing energy and beam power with the aim of gaining market shares from the booming electron beam cross-linking business, or creating smaller and simpler electron processors operating at voltages below 150 kV. The aim pursued with the latter approach certainly was to build low cost machines dedicated to printing and converting applications [10, 11, 14]. The hope was to obtain an investment cost level close to that of multi-lamp UV systems. Although this was not completely reached, the number of pilot and production EB installation increased from 320 in 1995 to 360 in 1997 [21].

However, cross-linking became an important market segment and EB curing of printing inks declined. On the other hand, a domain of EB curing, that is thick, pigmented and weatherstable coatings for outdoor applications, experienced considerable growth.

3. Low energy scanned beam electron accelerators

The most frequently used industrial low energy electron scanner is the one applied by Polymer Physik [22] (now Electron Crosslinking AB, Halmstad, Sweden). After its introduction in 1972, more than 60 R&D, pilot and industrial units were sold until 1997. The accelerator shown in Figure 10 uses a classical triode system as electron gun, a single acceleration gap, a beam focussing and a deflection system. The acceleration voltage ranges from 150 to 300 kV depending on the ultimate application.

The electron gun consists of a spiral-shaped tungsten cathode and a Wehnelt cylinder. The Wehnelt cylinder and the anode do not only constitute the electrodes of the acceleration gap, but also form an electron optical assembly to control and shape the electron beam. Linear current signals with repetition frequencies of about 800 Hz are employed to deflect the electron beam horizontally and vertically over the exit window plane. For maximum beam power, the scanner can be equipped with two cathodes. In that case the exit window has a width (in conveyor direction) of 220 mm instead of 100 mm for the single-cathode standard unit. The exit window is designed to have a large area and to enable effective cooling of the 12 to 15 μm thick titanium foil.

The general operation characteristics of Polymer Physik accelerators are given in Table I.

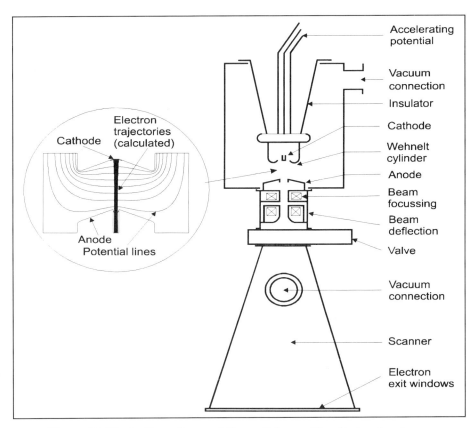

Figure 10 Single Stage Scanned Beam Polymer Physik Accelerator

Table I Operation Characteristics of Polymer Physik Accelerators

Acceleration Voltage	150 - 300 kV
Electron Current (per cathode)	200 mA max
Working Width	300 - 2000 mm
Electron Current per cm	3.2 mA/cm (max)
Web speed at 10 kGy	800 m/min (max)
Dose Variation Cross Web	± 5% max

The Japanese company Nissin High Voltage also offers electron scanners. The lowest possible acceleration voltage is given as 300 kV. The preferred voltage range of the scanned beam unit is 300 - 500 kV. Beam powers up to 65 kW are possible at a maximum beam width of 1200 mm.

4. Linear cathode electron accelerators

In 1970, Energy Science Inc. began to design a low energy electron accelerator which employed a cylindrical vacuum chamber in which a longitudinal heated tungsten filament was raised to a negative potential typically of 165 kV. The electrons emitted from the "linear" cathode were accelerated in a single step to the exit window. The exit window as part of the vacuum chamber was kept at ground potential. The electrons accelerated to the window penetrated the thin titanium foil and entered the process zone. This design principle allowed for a very compact size of the electron processor. Figure 11 shows a scheme of an Electrocurtain® processor. A linear cathode processor using a single filament is limited in dose-speed capacity to about 450 m/min at 10 kGy. Therefore, up to four filaments were grouped to a cathode assembly providing a dose-speed capacity of up to 1350 m/min at 10 kGy or a beam current of several hundred mA. Some general operating characteristics of Electrocurtain® processors are given in Table II.

VI – Electron Beam (EB) Curing Equipment

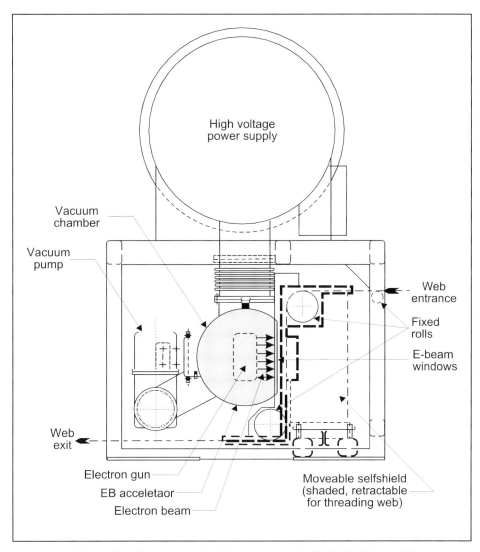

Figure 11 Electrocurtain® Processor with Selfshield Assembly

Table II General Operating Characteristics of Electrocurtain® Processors

Acceleration Voltage	150 - 300 kV
Working Width	150 - 2000 mm
Web speed at 10 kGy	1350 m/min (max)
Dose Variation Cross Web	± 10 % max

The processor is normally supplied with a Selfshield® product handling system. The design of the X-ray protection and product handling system has to be adapted to ultimate process demands. However, the block diagram of linear cathode electron processor control, as given in Figure 12, is in many respects typical of all types of low energy electron processors. Control is provided by programmable logic controllers (PLC) which are flexible and upgradable. Addition of man-machine interfaces (MMI) based on personal computers and common software systems open up possibilities not only for control and interlocks but also for history operation, communication and archiving. If the system can be upgraded to enable monitoring of real time signals, remote servicing from anywhere in the world becomes possible. The only equipment needed by a service engineer would be a telephone data line and a personal computer [21].

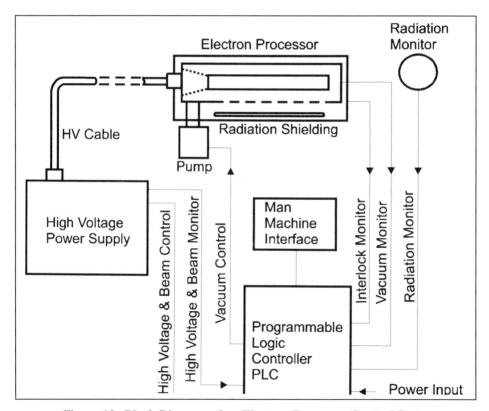

Figure 12 Block Diagram of an Electron Processor Control System

Following the trend to low energy, low cost processors, low energy linear cathode accelerators were developed and appeared on the electron processor market in the mid 1990s [9, 10]. These electron processors can be described as vacuum diodes operating in the current saturation mode, i.e. the cathode is a directly heated tungsten filament without a control grid. The electrons emitted from the cathode are shaped into electron beam bundles by a forming electrode and accelerated to the exit window. As an example, the configuration of an EBOGEN processor manufactured by IGM Robotersysteme, München, Germany [23], is shown schematically in Figure 13.

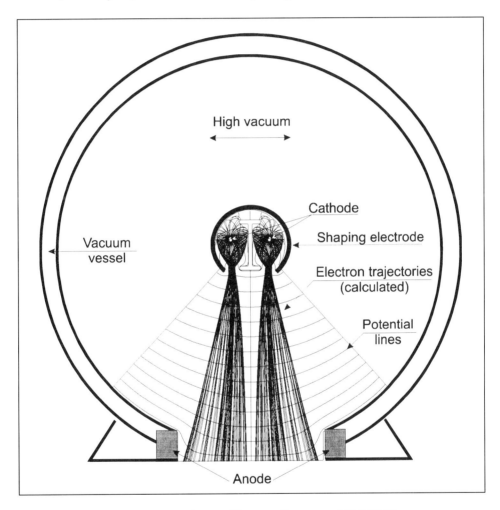

Figure 13 Low Energy Electron Processor EBOGEN

The diode configuration ensures that practically all electrons emitted are accelerated. Cathode heating can be accomplished through the high voltage cable. This allows moderate electron currents at low heating power. An open diode configuration forms a very simple electron optical system, but has the drawback that all emission in inhomogeneities of the heated tungsten wire are transferred to the accelerated beam illuminating the exit window. This leads to a dose variation across the web of up to 10 - 15%. The dose non-uniformity is lower for systems using two or more cathode filaments and especially designed cathode exit slits. Figure 13 is an example of a two filament "closed" electron optical configuration.

General operating characteristics of EBOGEN processors are given in Table III.

Table III General Operating Characteristic s of EBOGEN Processors

Acceleration Voltage	120 - 150 kV
Electron Current	120 mA (max)
Working Width	300- 1400 mm
Electron Current per cm	1.5 mA/cm (max)
Web speed at 10 kGy	450 m/min (max)
Dose Variation Cross Web	± 12 % max

5. Multi-filament Linear Cathode Electron Accelerators

In 1980 RPC industries, under the trade name BROADBEAM $^{(TM)}$ introduced an electron processor using a multiple emitter assembly with heated tungsten filaments placed in parallel to product direction. This design principle enabled the generation of large area electron beams with a uniform current density. As shown in Figure 14, the cathode filaments are mounted in parallel to the product direction. The beam current of the BROADBEAM $^{(TM)}$ processor is controlled by molybdenum grids held at a common potential. A planar screen grid is placed below the control grids. In the field-free region between the grids, a highly uniform electron density is generated prior to acceleration. Acceleration takes place between the planar screen grid and the grounded window.

VI – Electron Beam (EB) Curing Equipment

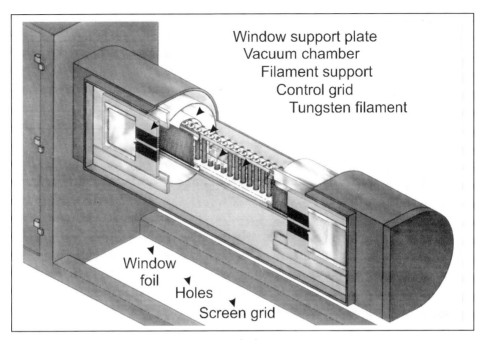

Figure 14 BROADBEAM (TM) Processor (schematically)

General operating characteristics of BROADBEAM (TM) processors are given in Table IV.

Table IV General Operating Characteristics of BROADBEAM (TM) Processors

Acceleration Voltage	150 - 300 kV
Working Width	300- 2500 mm
Web speed at 10 kGy	1500 m/min (max)
Dose Variation Cross Web	± 5 % max

In 1993, ESI announced the market introduction [14] of a new series of low energy, lower cost electron processors named Electrocure (TM). The use of a new window design combined with a multi-filament cathode arrangement made it possible to increase the dose-speed capacity of a 150 keV processor to 1050 m/min at 10 kGy. An even more powerful 120 keV, 200 kW Electrocure (TM) processor was installed in 1996 to cure coatings on 1650 mm wide web.

6. Recent developments in electron beam curing equipment

In electron beam curing technology development efforts are directed
- to achieve lower equipment costs,
- to build more compact and light-weight EB units and
- to avoid nitrogen inerting.

All these aims can be approached by lowering the electron acceleration voltage. Lower accelerator voltage means a less expensive power supply and a smaller size of accelerator head and x-ray shielding. At lower energy the penetration depth of the electrons into the substrate decreases. As a result, the surface dose increases (see Figure 4). Thus, ink and thin coatings can be cured at higher efficiency and speed. The high surface dose rate leads to the higher radical formation rates thus reducing oxygen inhibition.

As mentioned above, the design of the beam exit window determines the minimum electron energy. Based on an improved titanium foil window design, ESI now offers "Microbeam" EB units with typically 100 kV acceleration voltage and a maximum web speed of 660 m/min at 10 kGy [24].

Another new technical approach builds upon a sealed vacuum tube as electron generator which uses as beam exit a 2.5 µm thick 2 × 25 mm silica ceramic window. Using this innovative window design, now the electron beam tube can operate at 30 to 70 kV [25].

The smallest unit is a single Min EB vacuum tube that emits 1 to 2 mA. A 5 EB tube module contains 5 Min EB tubes in one housing and is used as basic building block for a so-called MTM curing module. The 5 EB tube modules can be stacked to achieve the desired beam width. Typical dimensions of a 25 cm wide shielded MTM module with idler rollers for web guidance are 130 × 45 × 60 cm. The modular design of the "Min EB" system is illustrated in Figure 15.

VI – Electron Beam (EB) Curing Equipment

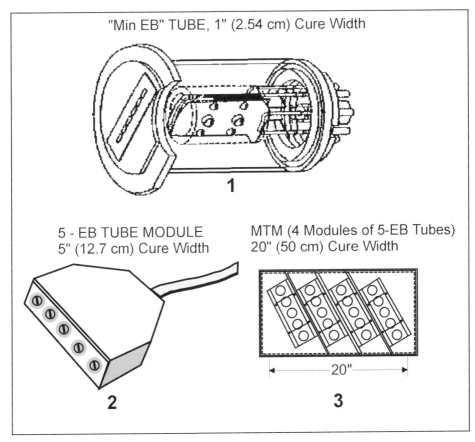

Figure 15 Modular Design of the "Min EB" system

This is an emerging technology that has to be proven on the market place with respect to coat, performance and reliability. It clearly competes with UV curing multilamp systems.

III. REFERENCES

1. A.H. Samuel, J.L.Magee, J.Chem.Phys. 21, 1080 (1953)
2. D.W. Clegg and A.A. Collyer (eds.), Irradiation Effects on Polymers, Elsevier, London (1991)
3. A. Singh and J. Silverman (eds.), Radiation Processing of Polymers, Hanser, Munich, Vienna, New York, Barcelona (1992)
4. R.Mehnert , Radiation Chemistry; Radiation Induced Polymerisation in Ullmann`s Encyclopedia of Industrial Chemistry, Vol. A 22, p. 471 (1993) , VCH Weinheim
5. P. Holl,E. Föll, Proceedings RadTech Europe `93, p.615 (1993)
6. E.P. Tripp , S.V. Nablo, Radiat. Phys. Chem. 14, 481 (1979),
7. W.A. Frutiger, S.V. Nablo, Radiat. Phys. Chem. 25,683(1985)
8. K. Mizusawa, T. Kimura, S. Uehara, Proceedings RadTech Asia'91, p.501 (1991)
9. R. Mehnert, P. Klenert, Radiat. Phys. Chem. 35, 645 (1990)
10. U. Schwab, Industrie-Lack 7/97 416 (1997)
11. P. Seifert, private communication
12. T.J. Menzenes, Radiat. Phys. Chem. 35, 52 (1990)
13. D. Meskan, A.F. Klein, Proceedings RadTech Europe '91, p.93 (1991)
14. P.M. Fletcher, Proceedings RadTech Europe'93, p. 637 (1993)
15. ICRU, Radiation Dosimetry, Electron Beams with Energies between 1 and 50 MeV, Report Nr. 35, Bethesda 1984
16. T. Tabata, R. Ito, Nucl. Sci. Eng. 53, 226 (1976)
17. R. Mehnert, Proceedings RadTech Europe '95, p. 1 (1995)
18. S.V. Nablo, J.R. Uglum, B.S. Qintal in, Nonpolluting Coatings and Coating Processes, eds. J.L. Gardon and J.W. Prane, p.173-193, Plenum Press, New York 1973
19. P. Holl, Industrie-Lackierbetrieb, 48,313 (1980) and 48,362(1980)
20. A.M. Rodriguez, W.T. Newcomb, Radiat. Phys. Chem.25, 617 (1985)
21. D.A. Meskan, A.F. Klein, Proceedings RadTech Europe '97, p. 114 (1997)
22. P. Holl, Radiat. Phys. Chem. 25, 665 (1985)
23. U. Schwab, Proceedings RadTech Europe`97, p. 105 (1997)
24. U. Läuppi, private communication
25. J.I. Davis, G. Wakalopulos, Proceedings RadTech North America `96, p. 317 (1996)

CHAPTER VII

UV CURING TECHNOLOGY - UV CURING UNITS AND APPLICATION TECHNIQUES

I. UV Curing Units - General Design Principles

1. Design Requirements

A wide variety of UV curing applications is now available. Because of the increasing number of installations and the development of new applications, UV curing systems become ever more complex. Nowadays, UV curing systems have to meet the needs of special product options in curing coatings or inks on:
- two-dimensional substrates,
- cylindrically shaped parts,
- three-dimensional parts

and even in spotcure or robotics applications.

Unlike EB, curing equipment is available in numerous sizes, power, in-line and off-line, in combination with large coating or printing machines, together with alternative drying equipment, and as stand-alone units, there are some general guidelines available to provide a high probability of technical excellence with UV implementation.

It is the answer to the key question: Will the UV system be able to cure all the products desired, in high quality, which finally defines the design requirements? In this sense, the basic technical needs can be discussed according in terms of:
- cure efficiency
- heat management
- system reliability and
- suitable integration in the production line.

2. UV cure efficiency

Efficient cure means that in all applications cases conditions should be created that a minimal UV dose (measured in $mJcm^{-2}$ on the substrate surface) or an minimum UV exposure time can be applied to obtain the desired degree of cure.

Excess UV exposure is costly, can lead to photodegradation of the substrate material, can cause substrate overheating and unnecessary ozone emission. Therefore, conditions should be chosen which ensure that UV curing can be conducted at a minimal dose.

First, the reactivity of the coating or ink should be optimised with respect to the UV light source used. This can be achieved by using reactive binders and a photoinitiator system which closely matches the emission spectrum of the UV source. This is often the job of the formulator.

Second, the minimal UV exposure time necessary perfectly to cure a chosen formulation should be determined under conditions which, with respect to spectral output and UV irradiance, match as close as possible those of the desired application case. This can be done experimentally by using either a suitable UV lab unit or a time-resolved analytical method such as real-time IR spectroscopy. In the latter case, the photochemical dark reaction can be observed, i.e., monomer conversion continues to proceed after UV exposure has finished. Thus, an exposure time Δt_{exp} can be determined which is needed to reach the necessary degree of cure (see Chapter VIII, Figure 7) and which is usually much shorter than the total reaction time. When speaking about "cure efficiency", one has to keep in mind that the photoinduced dark reaction (often called „postcuring") can cause the main part of monomer conversion.

Once the UV exposure time Δt_{exp} needed to reach an appropriate cure level is known, the irradiance in the irradiation plane I_0 (measured in $mWcm^{-2}$) is the second important parameter, which determines the maximum product speed. Both, irradiance profile and the amplitude have to be taken into account. The irradiance profile should be determined using a UV irradiator unit of exactly the same type and geometry that will be used in the desired application. Lamp radiative power, reflector geometry and the distance between lamp and irradiation plane determine shape and amplitude of the irradiance profile.

As schematically shown in Figure 15 of Chapter VIII, in terms of product direction the irradiance profile defines a irradiation interval Δx. The product speed is v_s simply obtained by:

$$V_s = \Delta x / \Delta t_{exp} .$$

The following examples (see Figure 1 and 2) illustrate how the maximum product speed can be estimated for curing of a 1.5 µm thick web offset ink (magenta) by using a UV curing unit equipped with two low power 308 nm excimer lamps:

Using as analytical tool real-time ATR - FTIR spectroscopy with 11 ms time resolution, monochromatic 313 nm excitation (interference filtered from a medium pressure mercury lamp), an exposure time Δt_{exp} of 40 ms, an irradiance of 50 $mWcm^{-2}$ and nitrogen inerting, the conversion vs. time profile shown in Figure 1 was measured.

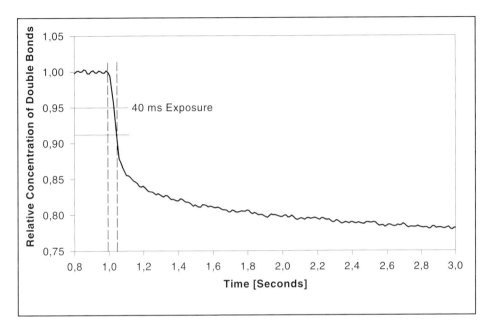

Figure 1 Web Offset UV Ink Magenta: Double Bond Conversion as Function of Time Measured After 40 ms Exposure with 313 nm Light, Irradiance 50 mWcm^{-2}

Within our time resolution an induction period is not observed. Immediately after the exposure the double bond conversion reaches 12%. This "prompt" double bond conversion is followed by the "dark" conversion: after one second the conversion increases to 22%.

Inspection of the cured ink on the diamond ATR crystal showed that the ink surface was solid but could be removed by a few MEK rubs. Therefore, in a following experiment the ink was irradiated by a series of exposure pulses. The pulse duration was reduced to 30 ms and the time delay between two pulses was set to 200 ms. The results of the multi-exposure experiment are shown in Figure 2.

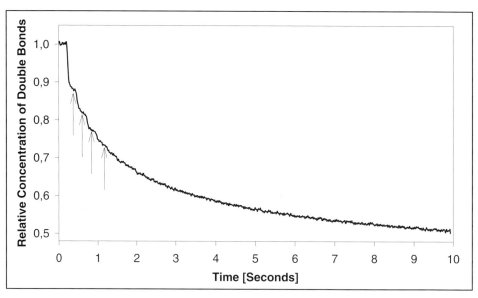

Figure 2 Web Offset UV Ink Magenta: Double Bond Conversion After Exposure with a Series of 30 ms Pulses of 313 nm Light, Delay Between Pulses 200 ms, Irradiance 50 mWcm^{-2}

The effect of the first and the second irradiation pulses can be seen as steps in the prompt double bond conversion, whereas the effect of the third and fourth is less pronounced and that of all following pulses disappears in the background noise.

Within the following 9 seconds a strong dark reaction can be observed. This is in accordance with the finding that the ink cured on the diamond surface is now much better MEK rub stable.

The following example illustrates, how the cure speed can be estimated from real-time FTIR results:
- Profile and amplitude of the irradiance created by the 308 nm lamp in the product plane (200 mm below the lamp) is measured using a CUV 270 silicon carbide diode which was calibrated by chemical actinometry (see Figure 7 of Chapter V). If the area of the irradiance profile is kept constant Δx and an averaged irradiance Im can be assumed as 5 cm and 200 mWcm^{-2}, respectively. Power and geometry of the excimer irradiator entirely correspond to an industrial UV curing unit for web offset printing which consists of two 308 nm lamps mounted on an inerting plate (see Figure 3).
- Assuming that 30 ms two-lamp exposure provides the desired degree of cure and that an increase in irradiance by a factor of 4 (from 50 to 200 mWcm^{-2}) leads to half of the exposure time, the maximum product speed is estimated as:

$$v_{prod} = \Delta x/\Delta t_{exp} = 0.05/0.015 = 3.3 \text{ ms}^{-1} = 200 \text{ mmin}^{-1}.$$

In practice, using the same ink and UV curing unit as shown in Figure 3 a product speed of at least 300 m/min is obtained. There are mainly two effects which prevent a better simulation of practical cure speeds: the uncertainties imposed by ink thickness differences and the long lasting dark reaction.

Especially for fast running curing processes, it is the dark reaction which mainly provides conversion.

Hence, to obtaining a high cure efficiency also means to create conditions for an effective dark reaction.

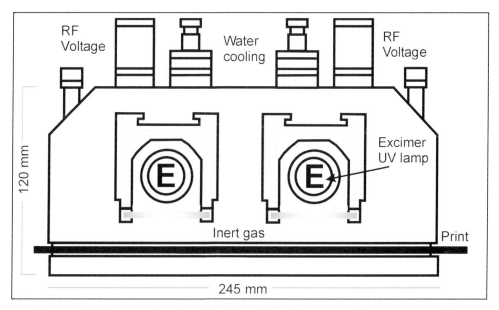

Figure 3 UV Curing Unit Used for Web Offset Printing of Heraeus Noblelight

Keeping in mind that the dark reaction is of great importance in UV curing, the concept of a "curing dose" needs critical evaluation. The curing dose D defined as the product of irradiance I_0 and exposure time Δt_{exp}

$$D = I_0 \times \Delta t_{exp}$$

only describes the contribution of prompt monomer conversion to total conversion. Therefore, UV dose measurements allow some relative process control but not the determination of the degree of cure.

Another important parameter affecting cure efficiency is the presence of oxygen.

During UV exposure monomer conversion starts after an induction period. For maximum cure efficiency, the induction period should be kept as short as possible. This implies that oxygen removal from the reaction zone, as usually done by nitrogen blanketing, is essential for obtaining highest cure efficiency.

3. Heat Management

All electrical input power needed to operate UV sources such as medium pressure mercury lamps and excimer lamps, either powered by AC, RF or microwave, is finally converted into heat.

However, infrared radiation emitted from the mercury vapour of polychromatic UV lamps and convected heat produced by infrared emission from the hot lamp jacket are the main source of heat in the UV curing process. In a medium pressure mercury lamp, the electrical power supplied to the plasma discharge is transformed to UV, visible, infrared radiation and convected heat. A comparison of the UV and infrared spectral radiances of different polychromatic and monochromatic UV lamps is given in Table I.

Table I UV and Infrared Emission from Polychromatic and Monochromatic UV Sources

Lamp Type and Power (Wcm^{-1})	Spectral Radiance UV (200-450 nm) Wcm^{-1}	Spectral Radiance IR (700-2500nm) Wcm^{-1}
H-Bulb 120 [1]	38	22
D-Bulb 120 [1]	75	22
H-Bulb 240 [1]	58	45
D-Bulb 240 [1]	120	45
Med.Pressure Hg Arc 120 [1]	48	50
308 nm Excimer Lamp 50	5	*
308 nm Excimer Lamp 240	65	**

* due to water-cooled lamp surface practically no convected heat
** convented heat about 25 % of the electric power

Radiation mainly penetrates the silica tube jacket, but part of the discharge energy will heat the tube, either by absorption of radiation or by non-radiative losses from the arc. In order to prevent mercury vapour condensation, a tube temperature between 900 and 1200 K has to be maintained for medium pressure mercury lamps. Air-cooling is the preferred method to keep the lamp temperature within the allowed range.

Decreasing the electrical power of the lamp below 40% without adapting the cooling conditions would result in a drastic loss of radiated power. Figure 4 shows a typical relationship between tube temperature and lamp power.

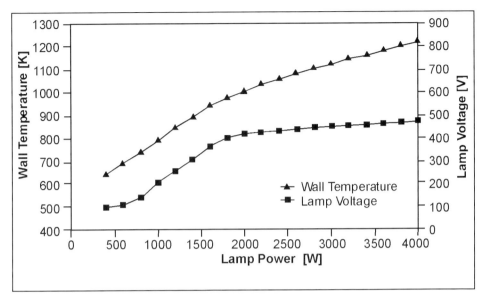

Figure 4 Tube Surface Temperature as a Function of Lamp Power

308 nm XeCl excimer lamp fillings do not contain any liquid components. Temperature dependent condensation processes do not play any role. Therefore, water cooling of the lamp jacket becomes possible. In dielectric barrier discharge driven excimer lamps the heat generated by the discharge can be completely transferred to the cooling water (see Figure 7). The surface temperature of the excimer lamp is only slightly higher than that of the environment. Infrared emission from the jacket can be neglected. Thus, the excimer lamp is cold and heat management is not necessary.

This does not apply to the microwave-powered 308 nm excimer lamp. The 2.45 GHz microwave frequency used is strongly absorbed by water. In this case, water cooling is excluded. Although air cooling is applied, the temperature of the tube remains comparable to that of microwave-powered mercury lamps.

For all medium pressure mercury lamps and also for microwave-powered excimer lamps, one primary source of infrared radiation is the hot fused silica lamp jacket.
The power P (in W) emitted by a surface area (A in m^2) of the temperature T (in K) and emissivity e can be described by the Stefan-Boltzmann law:

$$P = \sigma e A T^4, \qquad (1)$$

where $\sigma = 5{,}67 \times 10^{-8}$ $Wm^{-2}K^{-4}$ is a constant.

It can be readily seen from Stefan-Boltzmann's law that a reduction in heat generation can be achieved by decreasing the temperature and the emitter area. For medium pressure mercury lamps and microwave-powered excimer lamps, air cooling of the tube is commonly applied. Both measures show severe limitations: in the case of mercury lamps the tube temperature should not be decreased below 900 K. On the other hand, as the exchange of heat from air to solid is poor, the tube temperature of an excimer lamp remains too high. The second changeable parameter, the tube area, is defined by the design and cannot be altered in a wide range.

As already mentioned in Chapter III, IR reflection from an aluminium reflector surface can be drastically reduced by applying a series of dielectric absorbing layers on the aluminium substrate. Such a dichroic reflector is partially transparent for IR, while UV and visible radiation is mainly reflected. The reflector substrate is water cooled and forms a perfect sink for the IR energy. The reflector temperature is kept low. Reemission of heat is practically excluded.

For extremely heat sensitive applications, water IR filters are sometimes used. Such water filters can be either manufactured as dual walled quartz jackets or sealed flat plates. In all water filter applications, significant UV losses caused by multiple reflections on the quartz surfaces, have to be tolerated.

Another way to prevent IR irradiation of the substrate to be treated is to place a "hot" mirror between the lamp and the substrate. This is a quartz plate with a series of very thin dielectric layers which reflect IR and transmit shorter wavelengths. Again the loss of IR radiation is accompanied by considerable UV losses.

Figure 5 schematically illustrates the different methods used to control IR radiation from medium pressure mercury lamps.

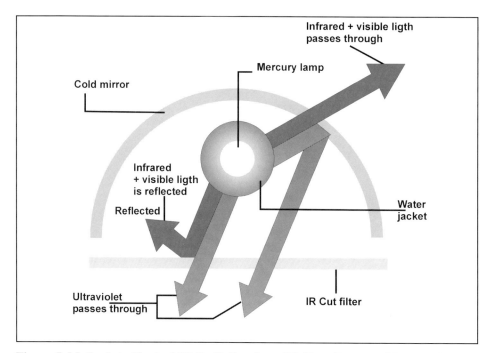

Figure 5 Methods to Control IR Radiation from Medium Pressure Mercury Lamps

Heat management of UV sources means strong reduction in IR power, which is emitted in the direction of the substrate plane, at constant or only slightly decreased UV output.

Additionally, a wide range of power control should make it possible exactly to select the power needed for the individual UV curing process.

Heat management for coating and substrate means to keep the temperature rise of the absorber in an acceptable range. The temperature rise of an IR absorber is governed by four factors:

- IR irradiance,
- IR absorption of the acceptor,
- its thermal conductivity and
- thermal convection.

IR irradiance in the substrate plane is determined by the power and nature of the UV source and by the focus conditions. If IR irradiance is too high, power reduction or defocussing is a simple way to control the temperature rise.

On a molecular basis, IR absorption leads to vibrational excitation, i.e., the vibrational spectrum of the absorber determines its IR absorptivity. In general, clear films show lower absorption than pigmented ones. Surfaces with a dark appearance are strong IR absorbers.

All coatings, inks and substrates show a certain thermal conductivity. Heat transfer to a sink is proportional to a material constant, the specific thermal conductivity, and the temperature difference between the heat source and the sink. Substrates of low thermal conductivity are often passed over a water-cooled drum or metal plate. The tight contact to the cold metal surface enables efficient cooling. On the other hand, coating of metal cans using cationic curable formulations is an example, where heating of the coating surface supports the cure process.

Often thermal convection is also used as a means to control the temperature of the substrate. Here, a turbulent flow of air is applied to cool the surface.

Heat management also means instantaneous stop of irradiation at a standing substrate. Fast operating water-cooled shutters are usually applied to prevent substrate deterioration by overheating. When the shutter closes, the lamp power is automatically reduced to stand-by mode.

To demonstrate the heat management by means of a practical example, a UV curing system developed for web offset is chosen [3].

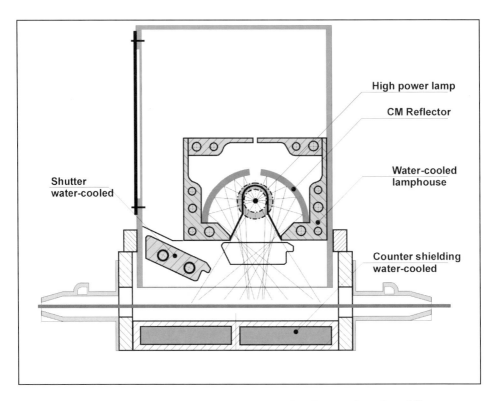

**Figure 6 Heat Management in a UV Curing System for Web Offset:
BLK System of IST UV Technology**

The output power of the 200 Wcm^{-1} medium pressure mercury lamp is steplessly controlled between 40 and 100% and slaved to the printing speed. Reduction of IR power is achieved by a dichrioc "cold mirror" (CM) reflector which is coupled to a water-cooled lamp housing as heat sink. The shutter is also water-cooled. Water-cooled shielding plates absorb radiation penetrating the substrate and serve as an additional heat sink.

An air exhaust system using charcoal ozone and dust filters not only prevents deposits of dust, volatiles etc. on the lamp but also provides convection cooling.

On the other hand, heat management can be limited to lamp water-cooling if a dielectric barrier discharge driven excimer lamp is applied. Figure 7 shows a UV curing system based on two 308 nm excimer lamps, which is also used in web offset printing [4]. Additional heat sinks are not necessary but nitrogen inerting ensures high cure efficiency.

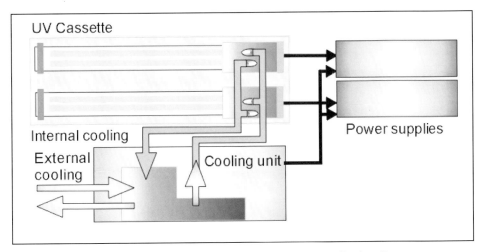

Figure 7 Heat Management in a UV Excimer Curing System

Finally it should be pointed out that nitrogen inerting of the irradiation zone is also an important means to keep the temperature rise low in the coating, the ink and the substrate. In comparison with UV curing under air, lamp power can be decreased by at least a factor of two with nitrogen blanketing. The resulting decrease in temperature can be significant.

4. System Reliability

UV curing system reliability does not just mean minimum down-time, it also implies perfect compliance with UV curing process quality. The skill of the UV equipment manufacturer is important for

- optimal lamp integration in the system,
- applying a suitable power supply and control unit,
- choosing the maximum power capability of the system to be at least 25% higher than needed for the process,
- creating the optimum cooling, inerting or gas exhaust conditions,
- providing permanent UV process control,
- providing easy and quick access to replacement parts (lamps, reflectors), and
- comfortable maintenance and service conditions.

Unless lamp life is limited by physical factors (see Chapter 3), the user of UV curing units containing medium pressure mercury lamps can do a lot to approach the theoretical lamp lifetime even under practical curing conditions:

The lamp -

- should not be frequently switched on and off, whereas stand-by operation has no negative effect,
- should be switched off only in intervals expected to be longer than 30 min,
- must not be touched with bare hands at the quartz tube,
- should be regularly cleaned with alcohol,
- should not be brought into contact with dusty or solvent-containing air, ink mists, coating volatiles etc.

The first two points mentioned above have to be modified if electrodeless microwave-powered units are used. These lamps practically have an instant on and off capability, which together with fast power control eliminates the need for shutters.

The same applies to dielectric barrier-driven excimer lamps. They may even be operated in a microsecond switching mode.

Chapter III has already discussed different technical solutions for lamp power control. The recent development of UV sensors, which are very stable against ageing, opens up new opportunities for permanent monitoring of UV intensity. As the fading of UV sensor sensitivity is much lower than lamp ageing, the sensor signal can be used to control the lamp power. Thus, permanent UV process control is achieved. Figure 8 of Chapter III shows a technical solution of UV process control.

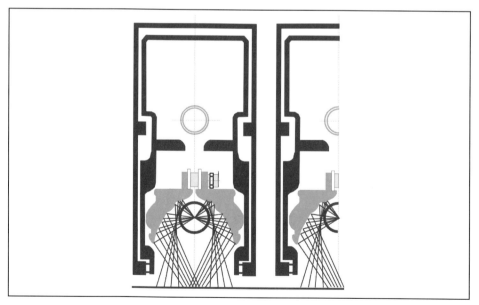

Figure 8 UV Irradiator Assembly Used in Offset Printing

To enable easy maintenance and service, durable steel or aluminium irradiator cassettes have been designed. They consist of the cooled single or multi-lamp housing, aluminium or dichroic reflectors, cooled shutters, an exhaust or inerting capability etc. These cassettes can be easily assembled to form larger UV irradiation units. Figures 8 and 9 show typical examples of UV irraditor assemblies.

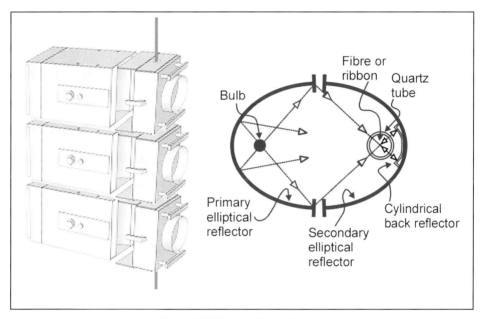

Figure 9 UV Irradiator Cassette

5. Integration in a production line

The UV curing system should form an integral part of the production line. Special attention should be paid to:
- master-slave operation of machine and UV curing unit in all states of operation: stand-by, start and production,
- status monitoring of electric power, shutter and reflector operation, temperature, cooling conditions,
- easy maintenance and servicing,
- minimum effect of the UV curing unit on machine parts: no heating, no ozone induced corrosion,
- safe UV radiation protection and ozone removal.

These are only a few general guidelines for the integration of UV curing units in production lines. In practice, an individual solution has to be found for every application case. Therefore, the next part of this chapter will discuss the design of UV curing units for different classes of typical applications in more detail.

II. UV CURING OF COATINGS AND INKS ON TWO-DIMENSIONAL SUBSTRATES

1. Curing of coatings or inks on flat, rigid substrates

In recent years, processors for UV curing of both functional and decorative coatings on flat, rigid substrates such as particle board, medium and high density fibreboard, natural wood, wood veneers, polycarbonate, polymethylmethacrylate, paper board and metal sheets have encompassed the parameters necessary to achieve a high cure quality and productivity.

(a) Wood finishing

UV curing units are essential parts of a finishing line, which in general consists of the following main components [5,6]:

- infeed,
- substrate pretreatment,
- first coating station,
- first processor for UV curing,
- treatment of the first coating,
- second coating station,
- second processor for UV curing,
- if necessary, repetition of the procedure,
- final coating,
- final curing,
- outfeed.

In general, the finishing operation begins with sizing and cleaning the substrate. For porous substrates, fillers are used to fill the pores and prevent the subsequent coating from penetrating into the substrate. To cover surface imperfections, thin overlay materials are laminated on the substrate in some cases. As a second layer, a pigmented colour coating can be applied, followed by a sealer and the top coat. It is the finishing schedule that determines the machine configuration for any finishing line. A typical finishing schedule for natural wood products such as furniture parts is given in Figure 10.

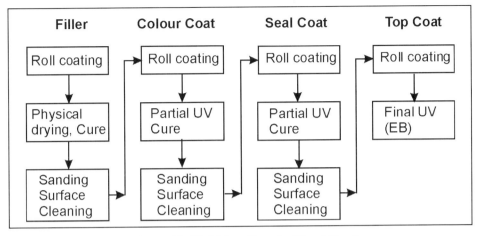

Figure 10 Finishing Steps in Wood Coating

Figure 10 gives roll coating as an example of a coating procedure. Although roll coating is often applied, curtain coating and even spray coating can be used in industrial finishing lines. Roll coating (maximum coating weight 40 gm^{-2}) and curtain coating (minimum coating weight 80 gm^{-2}) are the most efficient and economical ways to apply UV coatings, but they are limited to flat surfaces. To coat edges or three-dimensional parts automatic or manual spray application has to be used. Automatic multi-gun spray systems are available which allow reasonable product speeds up to about 10 mmin^{-1}. Overspray coating material is deposited on a conveyor belt from which it can be removed by a rotating doctor blade and recovered for reuse.

As a special coating procedure vacuum coating is recommended, e.g. for profiled medium density fibre board parts [7].

For clear coat wood finishing a decoration step is sometimes included in the process. Thermal transfer from printed polymer foils is one way of simple high-quality decoration. The transfer procedure, however, needs some time and results in limited product speed.

When applying several coaters and curing stations in tandem, partial curing between each coat means to expose the coat to a UV dose just enough to allow sanding but not to reach full through cure. This cure control allows for proper intercoat adhesion and film flexibility.

Final curing should provide proper through curing of the multi-coat system. Due to inadequate UV penetration, this could pose problems for thick pigmented coatings. In such cases, electron beam curing has to be preferred as the final curing technique.

A typical UV final curing station is schematically shown in Figure 11. It consists of one or more irradiator cassettes mounted on a reaction chamber, which contains a conveyor for product transport, a water cooled heat trap backing, different nozzles for

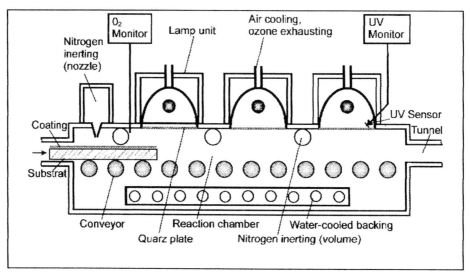

Figure 11 UV Curing Station for Coatings on Flat and Rigid Substrate

nitrogen inerting, a UV sensor and an outlet for monitoring the residual oxygen content, and provides appropriate UV radiation shielding. The stepless lamp control is slaved to the speed control of the conveyor and also reacts to the UV sensor level. Air-cooling and ozone filter are auxiliary units to operate the lamps. In this example, air- and water-cooled medium pressure arc lamps with shutters and cold reflectors are shown. If necessary, this lamp assembly can be replaced by a set of electrodeless lamps.

Quartz plates are placed between the lamp and the reaction chamber, which prevent interaction of air-cooling and inerting flow. Additionally, volatiles or dust do not come into contact with the lamp jacket but are deposited on the quartz surface, which is easily accessible for cleaning.

The benefits of UV curing under an inert atmosphere are also significant when using flat rigid substrates. Applying a sophisticated nozzle system as nitrogen infeed and a sensitive nitrogen control, a nearly laminar gas flow can be maintained in the reaction chamber during product transport. The residual oxygen concentration is monitored and can be used to automatically operate the nitrogen control system.

Typically two or three 120 Wcm^{-1} lamps are needed for wood finishing at line speeds of 30 $mmin^{-1}$ max.

UV curing of hard coatings on polymer sheets, metal decoration or screen printing on metal, polymers and electronic circuits are applications with a similar UV power demand. However, curing under air would require a UV power twice as high.

For fast and simple maintenance, the lamp housing and the upper part of the reaction chamber can be folded up pneumatically. This enables lamp and reflector servicing and cleaning of the conveyor zone.

(b) UV Laminating for print finishing

An alternative technique to protect and cover the sheet material is to laminate polymer film onto it [8].

When printed sheets are laminated with polymer films, this is called print finishing. One of the benefits of using a UV curable adhesive for laminating is that the bonding effect starts when the material is exposed to UV radiation. Once curing has started, a uniform and permanent bond is formed. Substrate and film are not affected by excessive heat or pressure.

A typical UV laminating line for print finishing is shown in Figure 12.
The printed sheets are cleaned from spray powder by brushing and hot calendering.
Using a three roll system the adhesive (3 - 5 gm^{-2}) is applied to the polymer film and laminated on the sheet. Directly integrated into the laminator unit is an automatic and line speed controlled sheet separation unit. The separated sheets pass a UV curing unit of the type described in Figure 11 where the adhesive is cured instantaneously. Finally, the finished laminate is fed into an automatic stacker. Using this UV laminator, a maximum of 10,000 sheets per hour can be produced up to a sheet size of 1020 x 1420 mm.

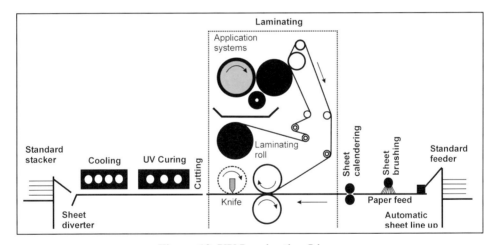

Figure 12 UV Laminating Line

(c) UV Screen printing

Another important application of the type of UV processors shown in Figure 11 is curing of screen prints [9].

In screen printing, ink is transferred through a stencil supported by a fine mesh. Metal meshes with up to 100 threads per cm form very fine pores. These pores are blocked up in the non-image areas and open in the image area. During printing the screen is flooded with ink. A squeegee is then drawn across it, forcing the ink through the pores and transferring it to the substrate.

The special benefits of UV screen printing if compared to the conventional process are [10]:

- no risk of ink drying in the screen,
- use of extremely fine meshes possible,
- fast drying means less substrate penetration of the ink.

2. Curing of printing inks on flexible substrates

(a) UV curing in offset lithographic printing: Web and sheet fed offset

In contrast to letterpress printing, where a relief printing image carrier is applied, lithography is a planographic process. This means that the image and non-image areas are in the same plane. To define image and non-image areas, lithographic printing plates are specially treated so that the image areas are hydrophilic whereas the non-image areas are hydrophobic. To cover the hydrophobic parts of the plate with a fine water film, the plate must be damped before inking. The hydrophobic areas cannot be covered by a closed water film but hold weakly bound tiny water droplets. When the ink roller is passed over the damped plate, it will push aside the water droplet and ink up the image areas.

However, the ink does not adhere to the areas covered by the water film.
The process is called offset lithography because the plate does not print directly on the substrate, but the image is first transferred (offset) to a rubber blanket and then to the substrate. Figure 13 shows the basic design of a lithographic printing unit which may be web or sheet fed. From the ink trough, ink is immersed by a manifold of ink rollers, which transfer the ink to the plate. The plate is clamped round the plate cylinder. At each revolution the plate is damped before inking. The blanket is clamped on the blanket cylinder. The blanket transfers the image to the substrate that passes between the blanket and the impression cylinder.

For web and sheet fed offset, two concepts of UV curing are used: final curing or a combination of interdeck and final curing. Printing wet-in-wet of up to four colours and subsequent drying is possible without significant loss in quality. However, interdeck UV curing prevents trapping of one colour on another and stops dot gain.

Figure 14 shows typical positions of UV curing processors in a web and sheet fed press. In web offset, simple technical solutions exist for nitrogen inerting of the UV curing zone.

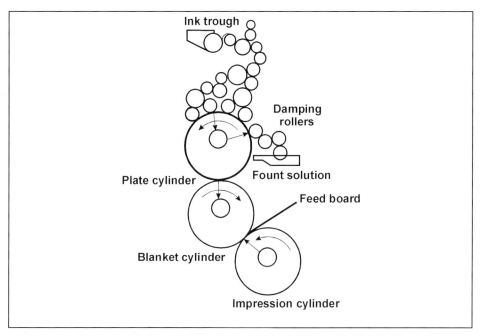

Figure 13 Design of an Offset Lithographic Printing Unit

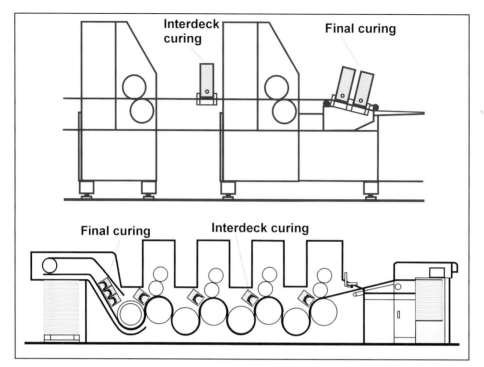

Figure 14 UV Curing Processors in Web and Sheet Fed Offset Presses

Excellent heat management is of special importance in UV curing on offset presses. UV processors as shown in Figure 3 and 7 are used in web offset, but also electrodeless microwave-powered lamps can be applied.

In UV curing of offset printing inks, the benefits offered by nitrogen inerting are quite obvious:

- curing speeds up to 400 mmin^{-1},
- lower energy consumption for curing ,
- less odour, volatiles and extractables from the prints,
- no ozone production.

Because impression and transfer cylinders must have clamps for sheet transporting, which act as air pumps, inerting poses problems for sheet fed presses. However, first technical solutions for inerting of UV curing units on sheet fed presses are now available.

(b) UV Curing in flexographic printing

Flexographic printing is a letterpress process that requires a relief printing image carrier. The area to be printed is raised above the non-printed area. When a flexible polymer printing plate is used, the letterpress process is called flexographic or flexoprinting. There are two basic types of flexopresses: machines consisting of several stand-alone printing units and CI presses, where the printing units are arranged circumferentially around a single central impression (CI) drum. When stand-alone printing units are used, UV curing in flexoprinting is quite similar to UV curing in web offset. However, implementation of UV curing on flexopresses not only leads to consequences for designing the UV curing processor but also for the printing unit [11,12].

The high viscosity of the UV-flexo inks in comparison to solvent- or water-based inks, causes problems when conventional doctor blades, aniloxes and plates are used. A possible technical solution of the problem is shown in Figure 15. The UV ink is pumped to a chamber doctor blade. Moderate pressure is applied to fill the cells of the anilox roller. In addition, special polymer plate material is used which shows good wetting properties for UV inks.

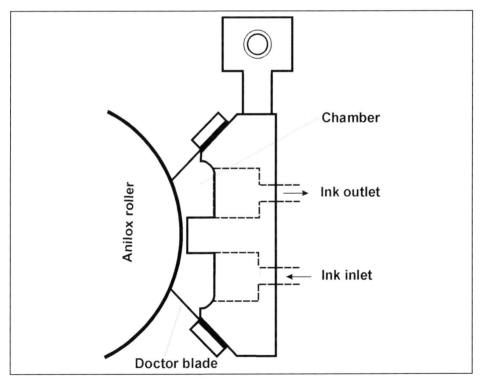

Figure 15 Ink Flow in a UV Flexoprinting Unit

In CI flexo presses, the UV curing units have to be placed between the printing units. Figure 16 shows a typical installation. It can be seen from Figure 16 that there is little space between the printing units, and the infrared energy emitted by the lamp is directly transferred to the drum. The drum is a high-precision cylinder exhibiting a concentricity tolerance of less than 5 µm. Therefore, the UV curing processor must not only be very compact in size but also as cold as possible. Water-cooled medium pressure arc lamps with good heat management and exact process control are one possible approach to a satisfactory technical solution. However, an additional cooling unit is required to keep the temperature of the drum within a few tenths of a degree and the drum within the concentricity limit. For a wide-web flexo press the additional cooling power can reach 100 kW and more.

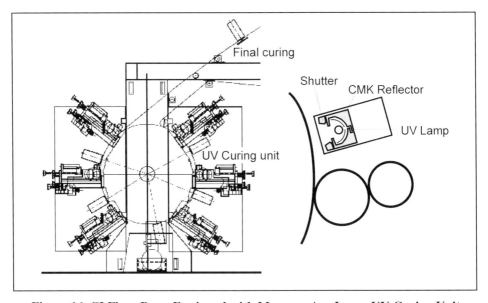

Figure 16 CI Flexo Press Equipped with Mercury Arc Lamp UV Curing Units

Using 2×200 Wcm^{-1} medium pressure mercury arc lamps in water-cooled housings as curing unit, printing speeds of 150-250 mmin^{-1} are now state of the art for radical inks. There are few curing problems with standard colours at a standard colour density and balance. However, the lower limit of the curing speed is caused by the colour white at a high opacity. The same applies for cationic inks, with the restriction that the curing speed is in the range between 100 and 180 mmin^{-1}.

Again inerting of the UV curing units in flexo printing is a way to:
- reduce the UV energy for curing,
- reduce the heat delivered to the CI drum and
- to avoid ozone, odour and volatiles.

Figure 17 shows another technical approach of UV curing processors for a CI flexo press. One or two 50 Wcm^{-1} 308 nm excimer lamps are placed on an inerting unit which uses the drum surface as a counter plate [13, 14]. A removable quartz plate is placed between the inerting chamber and the excimer lamp. No heat is transferred to the CI drum. Additional drum cooling is not necessary.

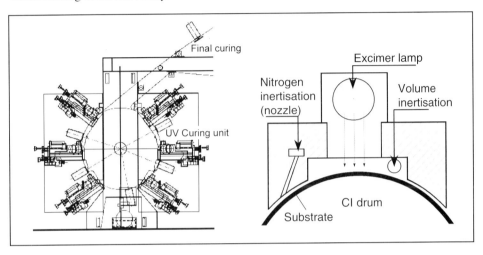

Figure 17 Inerted Excimer UV Curing Unit for CI Flexo Presses

When using low-powered 50 Wcm^{-1} excimer lamps, nitrogen inerting and tuning of the ink photoinitiator system to the excimer emission of 308 nm are necessary conditions to achieve reasonable curing speeds.

For radical standard inks at standard colour density and balance, cure speeds up to 200 mmin^{-1} can be obtained. As usual, highly opaque white ink is most difficult to dry. However, a powerful final curing station separated from the drum and placed on top of the press (see Figure 17) can help to solve the problem.

(c) CV Curing in gravure printing

In the gravure process, the image area is engraved into a cylinder in the form of small cells which become filled with ink. The ink is transferred to the substrate by passing it between the gravure and the impression cylinder. Although the gravure process is widely used in UV coating, there are practically no industrial applications of UV curing in gravure printing. One reason for this might be the difficulties encountered with precise filling and clearing of the cells, when using highly viscous UV curable inks.

UV curing on high-speed web gravure machines seems to be viable. UV curing processors similar to that used in web offset printing could be adapted to UV gravure printing.

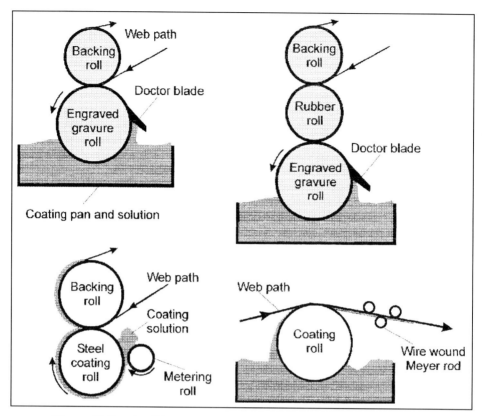

Figure 18 Coating Methods Used for Solventless UV Curable Fluids

Similarly to gravure printing, an engraved cylinder contains cells of different shapes, depths and distributions, which take up the coating fluid from the trough. A doctor blade wipes off the excess. The fluid left in the cells is then transferred to the substrate, either directly or via a rubber roll. To achieve easy filling of the reservoirs and stable transfer to the substrate, the fluids used in gravure coating should have a viscosity of less than 5 Pa·s and a low surface tension. Because the patterns of the gravure cylinder are „printed" on the substrate, measures are sometimes taken to level off the printed dots. A smoothing bar or indirect gravure via a rubber roll can be used to remove the printed pattern. In the gravure process, the coating weight is controlled by the cylinder pattern. The gravure method can be used for long runs and high precision.

3. UV curing of coatings on flexible substrates

(a) Coating methods

Formulations for UV curable coatings are mostly solventless 100% reactive systems. They are often highly viscous, and their rheology cannot be controlled in a simple manner. After transfer of the liquid coating to a substrate, UV curing occurs within a second or even less. The rough surface structure formed by the coating procedure is "frozen in" instantaneously. No surface "smoothening" takes place as known from solvent-based systems, where heat is applied to evaporate the solvent, thereby decreasing the surface roughness. This rough surface structure is a general peculiarity of many UV curable coatings and inks. For comparison Figure 19 shows the microstructure of an electron beam cured offset ink and a hot-air dried heat set ink.

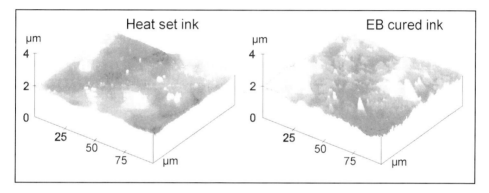

Figure 19 Surface Roughness of an EB Cured Offset Ink and a Hot-Air Dried Heat Set Ink

The selection of the appropriate coating method is of crucial importance for coating quality, functionality and economy. The coating method [15] has to selected according to:
- layer thickness (coat weight),
- coat viscosity and viscoelasticity,
- the coat weight accuracy required,
- coating speed and
- the substrate to be coated.

Figure 18 illustrates coating methods used for solventless UV curable liquids.

the web is transferred to the web. The coat weight can be exactly controlled by film splitting using a steel metering roll. The gap width between the metering and the application rolls controls the coat weight. In reverse roll coating all the fluid carried by the applicator is wiped off and picked up by the web moving in reverse direction. For higher precision and lower coat weights, four and five roll systems are often applied [16].

In rod coating, a wire-wound rod (Meyer bar) is used to remove the excess fluid from the moving substrate, thereby levelling the surface of the coating. This surface „shaping" procedure is often applied to achieve high gloss coatings. Usually the rod rotates in web direction or reverse to it. Rotation of the rod avoids collection of dirt particles on the rod. The coat weight is affected by the radius of the wire on the rod, fluid rheology, web speed and rotation of the rod. In a similar way, a blade or air knife is sometimes useful to shape the surface of a coating.

When surface shaping by a Meyer bar, by a blade or air knife gives insufficient surface smoothness, casting of the liquid coating against a highly polished chromium covered drum or a quartz cylinder is a relatively simple method to achieve a coating surface of nearly optical quality. The topology of the casting cylinder surface is reproduced by the coating brought into contact with the cylinder. Within a certain angle of rotation UV curing is accomplished, and the coated web is removed from the drum.

Table II summarises some coating methods and their characteristics.

Table II Coatings Methods and their Characteristics

Coating Method	Viscosity (Pa·s)	Coating Thickness (µm)	Coating Speed (mmin^{-1})
Roll Coating * Forward Reverse	0.2 -1 0.1-50	10-200 4-400	300 150
Gravure *	0.001-5	1-25	700
Rod (wire wound)*	0.02-1	5-50	250
Cast Coating	0.1-50	5-50	150

* Ref. [15] The numbers in this Table are only rough guidelines

(b) UV Curing of silicon release coatings

UV curable release coatings are based on polydimethylsiloxane oligomers, which have been functionalised with acrylate or epoxy groups. Silicone release coated papers and films provide "easy release" properties for a wide variety of pressure sensitive adhesives. Silicone release liners form a non-adhering surface upon which adhesive materials may be laminated.

To achieve perfect release properties, coat weights of about 0.7-1 gm^{-2} are sufficient. Multiroll reverse roll coating and offset gravure are the main coating procedures used in siliconisation. The type and the technical parameters of the UV curing system depend on the nature of the polysiloxane resin used in the siliconisation process. Silicone acrylates show radical cure whereas epoxysilicones form cationic curing systems.

The low coat weight and the transparency of silicone acrylate coatings should allow for easy UV curing. However, there is one peculiarity of silicone acrylates: the fluids have a relatively high oxygen saturation concentration, and oxygen diffusion is faster than e.g. in liquid acrylates. As a result, UV curing of silicone acrylates becomes very "sensitive" to oxygen. Nitrogen inerting of the curing zone up to an oxygen residual concentration of <50 ppm is a necessary condition for perfect curing. Therefore, inerted UV curing processors have to be applied. Because temperature sensitive substrates such as thin papers and polymer films are mainly used, and production speeds up to 600 mmin^{-1} are frequently desired, irradiator heat management and power control is of great significance. Inerted UV curing processors as shown i.e., in Figure 20 meet all requirements. Using 2×200 Wcm^{-1} medium pressure arc lamps, curing speeds up to 400 mmin^{-1} are easily obtained. Electrodeless medium pressure mercury lamps placed on an inerting chamber show similar cure efficiency and temperature effects.

As in the case of web offset printing, dielectric barrier discharge driven excimer lamps can also be applied as completely cold UV sources in siliconisation. The photoinitiator system has to be modified to give optimum radical yield upon 222 and 308 nm irradiation. Then the excimer UV curing processor described in Figure 3 of this Chapter can be used for curing silicone acrylates. A 2×50 Wcm^{-1} system allows a maximum curing speed of 150-250 mmin^{-1}.

Epoxy silicones as cationic curing systems are not oxygen sensitive. UV siliconisation can be done in air. For high speed applications the humidity in the reaction zone should be kept low. Therefore, dry air or nitrogen blanketing is sometimes used to establish stable curing conditions.

As in the case of UV curing of silicone acrylates, heat management is of crucial importance. Figure 20 shows an arrangement where UV irradiators were placed above a cooled drum which carries the substrate.

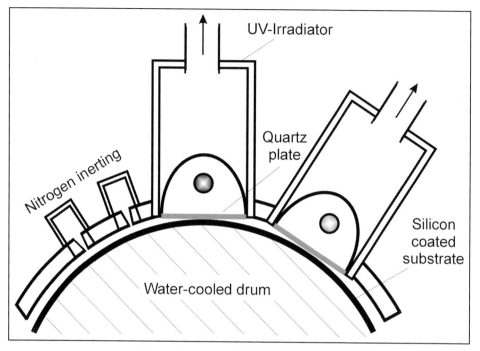

Figure 20 UV Curing in Siliconisation

(c) UV Curing of pressure sensitive adhesives

There are two types of radiation curable systems used as pressure sensitive adhesive (PSA) materials:

- systems which already have PSA properties but can be modified by radiation,
- systems with no inherent PSA properties before being irradiated [17].

In the first system, UV or EB radiation mainly induces cross-linking thereby modifying PSA characteristics such as shear value, peel strength, useful temperature range and resistance against solvents.

In the second system, radiation curing of acrylates containing tackifiers, acrylated polyesters or polyurethanes is performed up to a well-defined level which provides the desired PSA properties.

PSA formulations may show a wide viscosity range: from low viscosity liquids up to solids at temperatures below 80°C (hot-melts).Therefore, different application methods such as roll coating, screen printing and even a special type of extrusion coating are used. The coat weights are in the range of about 1 to 10 gm^{-2}.

In the past, mainly electron beam cross-linking of PSA`s was applied. Advantages of electron beam irradiation were simple electron beam dose control of the degree of cross-linking and no need for additives such as photoinitiators [18].

The development of UV monitoring equipment, which allows closed loop control of UV irradiance in the irradiation plane, removed the drawback of insufficient lamp control [19]. After addition of suitable photoinitiators to the PSA formulations, precise UV dose control enabled the desired adjustment of PSA characteristics.

A scheme of a stepless lamp control unit has already been shown in Figure 8 in Chapter III. The key element of the control system is the UV sensor. As it has to be placed in the irradiation zone, it must be cooled and kept clean during operation. The sensor is often installed below the water-cooled heat trap, which forms part of almost all UV curing processors, and is additionally air (or nitrogen) cooled.

The closed loop UV control system does not only provide UV irradiance but also computer-aided production control. Specific product and production data can be collected on an operation sheet. Thus, the mode of operation of the UV curing processor and the coater can be preset and controlled during operation.

(d) UV Curing of lacquers, varnishes and paints

UV curing of clear lacquers and varnishes is an important application of radiation curing. With respect to UV curing conditions, this class of coatings includes:

- clear overprint varnishes,
- finish coatings for PVC floor coverings,
- finish coatings for paper or film applied as laminates in wood decoration,
- transparent functional coatings providing special characteristics such as high abrasive resistance, gas barrier, conductivity, chemical resistance etc.

Coat weights of clear varnishes and lacquers vary from 2-5 gm^{-2} for typical overprint applications, and up to 10-100 gm^{-2} for finish and functional coatings. Therefore, coating methods such as gravure and reverse roll coating are used in most cases.

Because of the absorption of UV light by pigments, UV curing of paints suffers from inherent difficulties. To obtain the desired coverage, a paint weight of 50 gm^{-2} and often more has to be used. Additionally, a short exposure time is desired. The pot life of the paint should be acceptable and the paint should also show resistance against photoageing. To meet all these conditions, electron beam curing is the method of choice in curing thick pigmented coatings.

However, special photoinitiator combinations, which provide surface cure as well as through cure by photobleaching have been recommended [21] to allow reasonable cure speeds even at paint stabilisation against photodegeneration.

Thus, UV curing of paints with coat weights from 30-100 gm^{-2}, depending strongly upon the nature and concentration of the pigment becomes industrially feasible.

For UV curing of clear varnishes air- or water-cooled multilamp systems are commonly used which work without nitrogen blanketing. Especially for finishing plastic flooring materials up to 4 m wide production lines have to be equipped with UV curing units.

Single 4 m wide lamps are now available. This avoids non-homogeneous irradiance in the lamp overlap range. Metal halide lamps and lamp combinations are often employed to maintain a constant degree of gloss when applying matt finishes. Sometimes power-controlled pre-cure units are recommended for gloss adjustment. However, physical matting by means of 172 nm excimer radiation is also promising.

It is still difficult to UV cure thick pigmented coatings. To achieve the desired quality of cure, UV curing of paints and opaque coatings should be done under nitrogen at a high peak irradiance. A suitable choice of the lamp emission spectrum and the photoinitiator system is of crucial importance here.

(e) UV Curing in cast coating

Cast coating is a special method used to generate surface structures of the coating. Often the surface of a drum is used as "master", and "replication" can be done by curing the coating when in contact with the drum surface.

Figure 21 shows the principle of this procedure. When a metal drum is used, UV or electrons have to penetrate the substrate. For opaque substrates a quartz drum can be applied, where the UV lamp is placed inside. Cast coating is not limited to special coating formulations. In practice, residual-free removing from the drum can be achieved by simple formulation measures.

Cast coating against a perfectly polished drum is the only means to reach a microroughness of the coating surface which fulfils optical demands.

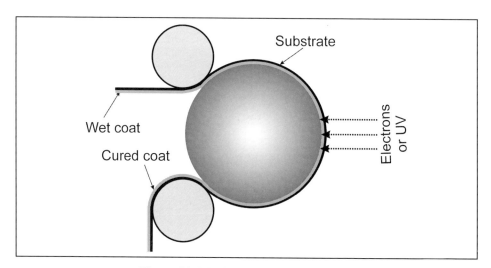

Figure 21 UV Curing in Cast coating

III. UV CURING OF INKS AND COATINGS ON CYLINDRICALLY SHAPED PARTS

UV curing on cylindrically shaped parts basically means curing of inks or coatings on plastic cups, tubs, tubes or metal cans. In comparison to the conventional process, UV curing gives the benefit of a simplified process and running at a faster line speed. For UV curing, the economical trade-off is between higher material cost and lower operating cost. For two-piece metal cans, UV curing of rim coating offers cost advantages [22], whereas UV printing and curing of plastic cups is practically dominating the market.

Figure 22 shows a typical arrangement for printing and drying cups on a mandrel.

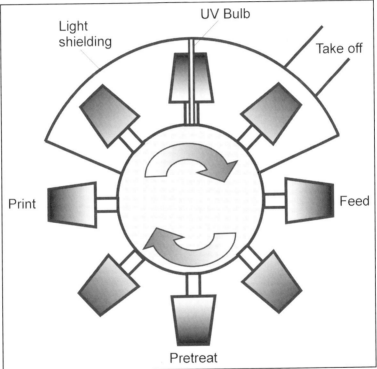

Figure 22 Arrangement for Printing and Drying Cups on a Mandrel

The most widely used printing process in cup printing is dry offset. This is a letterpress process where paste inks are rolled onto a relief type printing plate, transferred to a rubber blanket and printed onto the container. On its way through the printing machine the injection moulded or thermoformed plastic cup passes the following stations: the cup is blown onto a mandrel, corona or flame surface treated and fed to the printing station, where the entire picture is transferred from the blanket to the cup.

Compact and high-powered UV lamps allow curing of the rotating cup directly on the supporting mandrel. Curing times of 70 to 100 ms are common. Thus, highly reactive inks have to be used.

In the final step, the UV cured cups can be restacked immediately. Using UV curing, the overall size of the machine can be reduced by about 50%. The energy consumption decreased to less than 20% of the conventional process, and a production speed of 500 pieces per minute was achieved [23].

A rather similar technique can be used to UV cure inks and varnishes on lids, tubs and plastic squeeze tubes.

While radical inks and coatings are predominantly used in UV cup coating, a shift to cationic or hybrid inks can be observed in metal decoration. Cationic inks offer good adhesion to metal substrates. Additionally, abrasion resistance and elimination of skin irritation are important issues. Hybrid inks attempt to balance the characteristics of radical and cationic chemistries.

IV. UV CURING OF THREE-DIMENSIONAL PARTS

1. Three-dimensional (3D) UV curing

Design of equipment for 3D-UV curing must consider:
- size and shape of the part,
- method of coating and conveying,
- time and uniformity of UV exposure,
- required ventilation,
- shielding and interlocks of the finishing zone.

Because of their complicated shapes three-dimensional objects require special coating techniques. Spray, dip, flow and spin coating are procedures often used for 3D parts. In most 3D UV curing units the lamps are fixed and the coated parts pass in front of the UV lamps. The part should rotate two or three turns as it passes through the irradiation zone. This is shown for coated automotive headlamp components in Figure 23 [24].

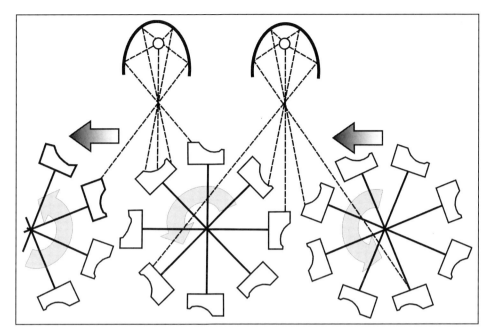

Figure 23 Translate and Revolve Technique in UV Curing of 3D Parts

A UV curing test plant recently built is equipped with eight microwave-powered lamps that can be moved by servo motors [25]. Automatic control allows for reproducible positioning of the lamps, and it is planned to move the lamps along with the parts to be cured. Thus, the required number of lamps, their positioning and the required conveyor speed can be adjusted to give optimum curing conditions.

Finishing of furniture parts is another well-established application of 3D UV curing. Overhead conveyors offer the greatest handling versatility in this case.

The UV irradiation chamber needs ventilation to dissipate heat and to remove ozone and volatiles. As a rough guideline, 3 D chamber must exhaust 8 $m^3 min^{-1}$ air per kW lamp power.

Shielding of the 3D chamber must be designed in a way to eliminate direct and first scattered light, and electrical interlocks and safeguards must be provided to prevent access to the chamber during irradiation.

2. Spot curing

Providing a spot of powerful UV light is an important method to cure adhesives and dental composites. Current dental polymer curing systems use 35 to 75 W tungsten filament quartz halogen tubes for fast curing of dental composites. An irradiance of 500 $mWcm^{-2}$ over a wavelength range from 400 to 500 nm can be obtained which cures dental composites within 30 seconds. For curing of adhesives, irradiances up to 2 Wcm^{-2} can be generated by a 100 W short arc mercury lamp coupled to a UV transmitting wave guide [26]. The flexible wave guide can be combined with a robotic mechanism thus providing another approach to curing of complicated surfaces.

V. UV MATTING OF COATINGS

The impression of a matt film surface is created by diffuse reflection of light from a micro-rough surface. Such a structure is obtained either by incorporating a matting agent into the coating or by changing the surface topology through shrinkage during polymerisation. Synthetic amorphous silicas are widely used as matting (or flattening) agents because of their porous nature and size distribution. In solvent-containing coatings matting poses no serious problem, because after evaporation of the solvent the contours of the particles on the surface form the micro-rough topology needed.

Radiation curable coatings are solventless. The shrinkage of the film is reduced to a minimum. Therefore, the matting efficiency of solvent free coatings is low. The matting effect is only caused by particles, which are already in the vicinity of the wet film surface. As a consequence, higher particle concentration and larger size particles must be used in UV curable coatings. This leads to serious problems with respect to thickening of the coating, sedimentation of the matting particles and curing reactivity.

As a result, the "chemical" matting procedure has to be replaced or at least modified by "physical" matting procedures.

One method to achieve physical matting is the so-called "dual cure process" [27]. When in a first step of UV irradiation, low irradiation doses are used the slow rate of cure causes the matting agent to migrate to the surface of the film. In a second irradiation step the film is through-cured.

This process was originally developed for polyester/styrene systems in wood finishing. It has to be modified when using acrylate based clear coats, e.g. for flooring or film coating applications. In this case, a first irradiation step induces curing in the bulk by means of metal halide lamps, whereas perfect surface cure is obtained in the following irradiation step by using mercury arc lamps. Additionally, the gloss level is controlled by the coating thickness, the temperature of the coating or substrate and the spacing of the UV lamps [28].

Another way to achieve matting by a dual cure procedure is based on cure inhibition by oxygen. In the first irradiation step a matt-appearing surface is created by surface cure inhibition. In a second irradiation step, either UV or EB, this surface structure is frozen in.

Another very interesting new method to generate low gloss finishes is the application of 172 (or 222) nm excimer radiation [29]. In a nitrogen atmosphere, 172 nm photons create surface cure in acrylate systems. The cured surface shows a topology as given in Figure 24 The deformed polymer layer floats on the liquid acrylate like a milk skin. The surface roughness leads to diffuse light reflection and creates a matt appearance of the surface. In the following irradiation step through-curing is initiated by UV or EB.

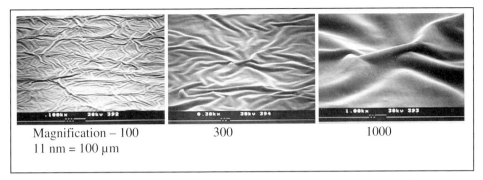

Magnification – 100   300   1000
11 nm = 100 µm

Figure 24 Surface of an Acrylate Coating after Exposure to 172 nm Excimer Radiation

VI. REFERENCES

1. R.W. Stowe, Proceedings RadTech`92 North America, p. 447 (1992)
2. A.B eying, Proceedings RadTech Europe `97, p. 77 (1997)
3. BLK-System of IST UV Technik, Nürtingen, Germany
4. Blue Light System of Heraeus, Hanau, Germany
5. A. Stranges, Proccedings RadTech `94 North America, p. 415 (1994)
6. D.T. Jones, Proccedings RadTech `94 North America, p.417 (1994)
7. P. Holl, E. Föll, Proceedings RadTech`93 Europe, p. 615 (1993)
8. R. Remund, Proceedings Radtech Europe`97, p. 386 (1997)
9. B. Blunden and J. Birkenshaw in "The Printing Ink Manual", Fourth Edition, R.H. Leach, C. Armstrong, J.F. Brown, M.J. Mackenzie, L-Randall, H.G. Smith (eds.), Chapman & Hall, London, 1991
10. M. Caza, Proceedings RadTech Europe`97, p. 392 (1997)
11. G. Bolte, Proceedings RadTech Europe´93, p. 265 (1993)
12. G. Bolte, Proceedings RadTech Europe`97,p. 379 (1997)
13. P. Klenert, J. Vogel, German Gebrauchsmuster 29613047.8
14. Patent WO/96/34700
15. E.D. Cohen, E.B. Gutoff (eds.): "Modern Coating and Drying Technology", VCH, Weinheim, 1992
16. T. Zimmermann, Proceedings RadTech Europe `93, p. 698 (1993)
17. R.W. Oemke, Proceedings RadTech `92 North America, p.229 (1992)
18. W. Karmann, Proceedings RadTech Europe`89, p. 595 (1989)
19. R. Müller, Proceedings RadTech Europe `97, p.191 (1997)
20. L.S imonin-Catilaz, J.P. Fouassier, Proceedings RadTech `94 North America, p.423 (1994)
21. A. Valet, D. Wostratzky, Proceedings RadTech`98 North America, p. 423 (1994)
22. D. Blake Proccedings RadTech Asia`97, p. 530 (1997)
23. K.F. Roesch, Proceedings RadTech Europe`93, p. 940 (1993)
24. R.W. Stowe, Proceedings RadTech`88 North America, p. 532 (1988)
25. W. Klein, M. Schneider,U. Strohbeck, Proceedings Radtech´98 North America, p.418 (1998)
26. R. Burga, Proceedings RadTech Asia`97, p. 192 (1997)
27. P.G. Garratt, Proceedings Radcure Europe`87, p. 10-23 (1987)
28. D. Skinner, Proceedings RadTech `96 North America, p. 458 (1996)
29. A. Roth,M. Honig, Proceedings RadTech`98 North America, p. 112 (1998)

CHAPTER VIII

RADIATION CURING TECHNOLOGY – UV CURING

I. DEGREE OF CURE AND CURE SPEED

1. Degree of cure

The definition of radiation curing as *"the fast transformation of 100% reactive, specially formulated liquids into solids by UV photons or electrons"* implies that during irradiation a conversion of reactive sites takes place until a solid network is obtained. In a strict sense, full cure is achieved when all reactive sites have been reacted. As a result, the degree of cure at a certain time (DC) is defined as:

$$DC = \frac{Number\ of\ reacted\ sites}{Total\ number\ of\ reactive\ sites.} \quad (1)$$

The degree of cure (at a certain point of time) DC can be given as a percentage if eqn.(1) is multiplied by 100%.

In radiation curing, typical reactive sites accessible to conversion are acrylate double bonds, vinyl groups or oxirane moieties. In a photoinitiated polymerisation reaction the reactive sites are consumed during chain initiation and propagation until all reactive sites have been reacted, or more realistically, until conversion has stopped. In the latter case some of the reactive sites are blocked by the solid network from being close enough to radicals or ions. A certain number of immobile "residual" reactive sites are generated. In particular, this is the case in curing of di-or multifunctional acrylates. Due to cross-linking, radiation cured formulations containing multi-functional acrylates form tight polymer networks showing a considerable amount of immobilised residual double bonds.

In most technical curing applications, reactivity of the formulation to be cured and irradiation conditions are chosen to achieve "full" conversion after less than one second. "Full" conversion means that at least in the timescale of seconds after irradiation, the concentration of reactive sites remains stable and cannot be further decreased by irradiation. If this steady-state concentration of reactive sites can be observed in a curing experiment, a second definition of the degree of cure can be applied:

$$DC = \frac{Number\ of\ reacted\ sites}{Total\ number\ of\ reactive\ sites\ -\ number\ of\ residual\ reactive\ sites.} \quad (2)$$

It should be pointed out that the number of residual reactive sites in a cured network is not a well-defined quantity but depends on monomer and oligomer functionality and on the mobility of the reactive sites within the polymer chain. Formulations with high reactive functionality show higher concentrations of residual reactive sites. In addition, the number of residual reactive sites is often affected by curing conditions such as photoinitiator content, temperature and irradiance.

If the timescale of curing observation is expanded from seconds to minutes, hours or even days it becomes obvious that conversion has not stopped after illumination but continues to proceed. Radicals which have been trapped in the solid matrix migrate to double bonds or recombine: a dark reaction or postcuring is taking place.

Postcuring is even more pronounced in the case of photo-initiated cationic polymerisation.

If postcuring is taken into account, definition (2) of the degree of cure has to be modified using the number of residual reactive sites remaining after postcuring has finished.

The two definitions of the degree of cure given above, which are based on conversion of reactive sites, are related to the reaction kinetics of cure. The meaning of cure can also be considered in more pragmatic terms. In the real world of curing technology most of the users prefer as definition of (sufficient degree of) cure that:

the cured product meets the needs of its function properties to be commercial   (3)

Apart from this pragmatic definition of cure, it is of great significance for a better understanding of the curing process to consider its kinetics and to evaluate the various techniques that are used to follow the kinetics of UV curing [1-8].

2. Evaluation of kinetic parameters in radical curing

(i) Polymerisation rate

Chapter I, Figure 10, summarised the different steps of photoinitiated radical polymerisation. Dividing the polymerisation kinetics into three main steps: initiation, chain propagation and termination, the following expression for the polymerisation rate vp has been derived:

$$v_p = k_p (k_t)^{-1/2} [M] (\Phi_i I_a)^{1/2} \qquad (4)$$

where k_p and k_t are propagation and termination rate constants, [M] is the monomer concentration, I_a is the number of photons absorbed per second and Φ_i is the yield of start radicals. In order to derive the polymerisation (= propagation) rate v_p, the following assumptions have to be made:

- monochromatic light is used, which is absorbed exclusively by the photoinitator,
- the absorption is small and homogeneous in the volume irradiated,
- when polymerisation proceeds, a stationary radical concentration is obtained,
- all polymer radicals show the same reactivity towards propagation and termination.

The number of photons absorbed per second in the sample to be cured can be expressed using Lambert-Beer's law

$$I_a = I_o (1 - \exp(-2.303\varepsilon[PI]l)), \qquad (5)$$

$$I_a = I_o \, 2.303 \, \varepsilon [PI] l. \tag{6}$$

In eqn. (6) ε (in l mol^{-1}cm^{-1}) is the molar decadic extinction coefficient of the photoinitiator at the irradiation wavelength, [PI] (in moll^{-1}) is the photoinitiator concentration and l (in cm) is the optical pathlength of light, corresponding to the thickness of the sample.

Using eqn. (5) the polymerisation rate v_p is obtained as:

$$v_p = k_p (kt)^{-1/2} [M] \, (\Phi_i \, I_o (1-\exp(-2.303\varepsilon[PI]l)))^{1/2} \tag{7}$$

In eqn.(7) I_o is given in mol l^{-1}s^{-1} or. According to eqn.(7) the polymerisation rate v_p can be plotted:

- versus the square root of I_o (at constant [PI]) or
- versus the square root of $(1-\exp(-2.303\varepsilon[PI]l)$ (at constant I_o).

(ii) Experimental determination of the degree of cure and the polymerisation rate

The polymerisation rate $v_p = -d[M]/dt$ can be deduced from the measured monomer conversion vs. time. In the case of photoinitiated acrylate polymerisation, conversion vs. time profiles can be measured by various methods, which will be outlined below in more detail.

To illustrate the theoretical considerations on UV curing kinetics given above by experimental examples, the following description uses experimental data which was obtained by real-time Fourier transform infrared spectroscopy (FT IR). Real-time (or time-resolved) FTIR spectroscopy is an elegant and powerful method to measure photoinduced monomer conversion as a function of time.

Figure 1 shows the scheme of the real-time FT IR apparatus used to obtain kinetic data on a model system consisting of tripropyleneglycol diacrylate (TPGDA) as monomer [M] and Irgacure 369 (see Table I) as photoinitiator [PI].

The infrared spectra were recorded in real time with a Biorad FTS 6000 FTIR spectrometer equipped with a MCT detector. The spectrometer is able to record up to 95 spectra per second at a spectral resolution of 16 cm^{-1}. A heatable single reflection diamond ATR device ("Golden Gate"; Graseby Specac) was used for sampling. Sample thickness and diameter are defined by a quartz plate into which a cylindrical hole of 4-μm depth and 15 mm cross-section was ion beam etched. The sample diameter considerably exceeds the diameter of active area of the diamond ATR crystal, which is 0.06 cm. The quartz plate covering the sample suppresses the contact of the sample with the surrounding air. Thus, oxygen diffusion into the sample is prevented during irradiation. In this case, radicals formed after irradiation exclusively react with oxygen dissolved in the liquid formulation.

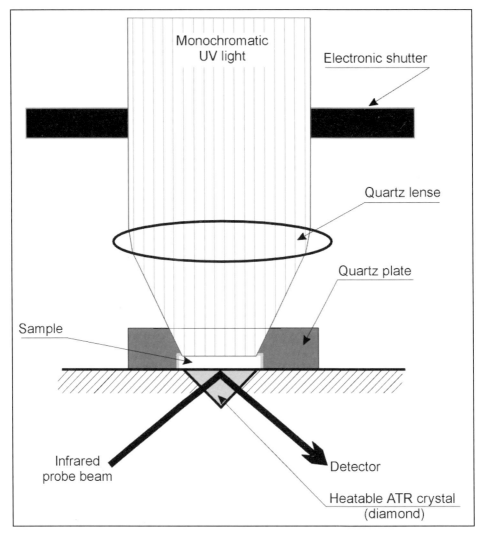

Figure 1 Scheme of an ATR Real-Time Fourier Transform Infrared (FT IR) Spectrometer

This instrument permits investigation at temperatures up to 200°C. Spectra with an excellent signal-to-noise ratio are obtained with this set-up even at the highest data recording rates.

UV-irradiation was performed with an Osram HBO 100 W mercury arc lamp. The light source is equipped with a water filter for blocking infrared radiation, neutral density filters to vary the light intensity, and a 313 nm metal interference filter to provide monochromatic light. The intensity of the incident UV light was measured by a UV radiometer using a SiC detector which is calibrated to radiation with a wavelength of 313 nm by chemical actinometry. The maximum irradiance measured at diamond crystal position was 100 mWcm^{-2}.

An electronic shutter directly driven by the spectrometer allows exact synchronisation between irradiation and spectra recording.

Typical double bond conversion vs. time profiles as measured for the model system TPGDA/IC 369 by irradiation with 313 nm photons at an irradiance of 48 mWcm^{-2} are shown in Figure 2. The photoinitiator IC 369 was chosen because of its good performance in formulations such as printing inks and its high exctinction coefficient ε (313nm) = 17.000 lmol^{-1}cm^{-1}. Table 1 summarises structure, triplet lifetime and quantum yield of α-scission for IC 369, parameters which will be used in the following examples.

Table1 Photoinitiator IC 369 (2-benzyl-2-dimethylamino-1-(4-morpholinophenyl)-butan-1-one): Structure and Kinetic Parameters [9]

Structure:	
Triplet lifetime:	1700 ns
Quantum yield of α-scission:	0.22

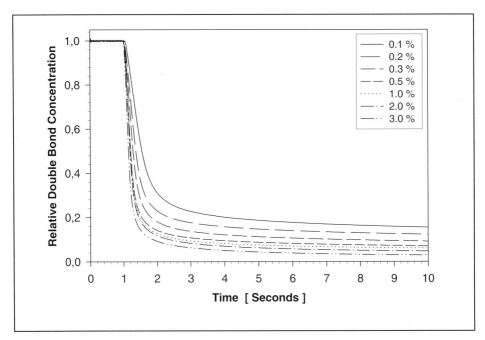

Figure 2 System TPGDA/IC 369: Double Bond Conversion as a Function of Time and Photoinitiator Concentration Measured during Irradiation with 313 nm Light, Irradiance 48 mWcm^{-2}

The conversion vs. time profiles were recorded after shutter opening at 1 s. Double bond conversion was followed by measuring the peak area centred at 810, 1190, 1410 or 1630 cm^{-1}. The decrease of the peak area reflects the double bond conversion and is exactly proportional to the degree of cure reached after a certain time. Immediately after shutter opening the conversion vs. time profiles show the typical steep decrease but level off after a few seconds.

The total degree of cure as given by definition (1) can be estimated from the (nearly) stationary residual concentration of double bonds reached in our experiment after 9 s. It amounts to 97% when 3 wt.-% of IC 369 is used and decreases to 84% at 0.1 wt.-% of photoinitiator. Figure 2 also illustrates the dependence of the residual double bond concentration on the photoinitiator content.

From the first derivative of the monomer conversion vs. time -d[M]/dt, the polymerisation rate v_p can be determined at any time. Taking into account that the total concentration of acrylate double bonds in TPGDA is 7.0 mol^{-1}, the polymerisation rate profiles shown in Figure 3 were derived from the conversion profiles of Figure 2.

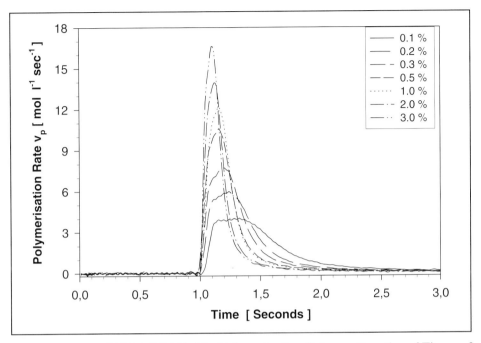

Figure 3 System TPGDA/IC 369: The Polymerisation Rate as a Function of Time and Photoinitiator Concentration (see Insert)

In the case of 3 wt.-% of photoinitiator, the maximum polymerisation rate is reached after about 100 ms corresponding to a degree of cure of 30%. The decrease in the polymerisation rate at higher conversion is due to both the progressive consumption of double bonds and the segmental immobilisation caused by the gel effect.

(iii) The polymerisation rate as a function of irradiance

As predicted by eqn. (4), a plot of the polymerisation rate v_p vs. square root of Ia, the number of photons absorbed per second, should give a straight line with a slope:

$$\alpha = k_p (k_t)^{-1/2} [M] \Phi_i^{1/2}. \tag{8}$$

Using our model system TPGDA/IC369 with 0.3 wt.-% of photoinitiator, the maximum rate v_p was measured as a function of the irradiance I_0. As shown in Figure 4, a plot of the maximum polymerisation rate $v_{p\,max}$ vs. square root of I_0 gives the straight line expected.

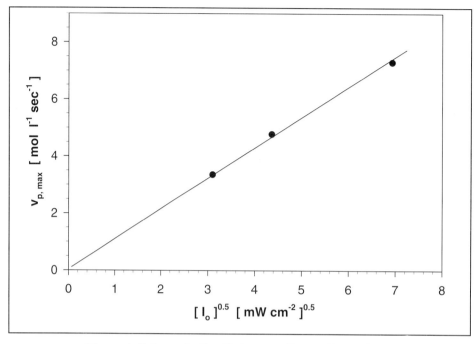

Figure 4 Polymerisation Rate v_p vs. Square Root of I_0

A small photoinitiator concentration was chosen in order to get small optical absorbance in the sample. As already mentioned above, monochromatic irradiation and small optical absorption (corresponding to low monomer conversion) are necessary conditions to apply eqn.(4).

(iv) The polymerisation rate as a function of photoinitiator concentration

Using Lambert-Beers law, the number of photons absorbed in the sample per second and unit area Ia can be calculated from the irradiance I_0, the photoinitiator concentration [PI] and the photoinitiator extinction coefficient ε at irradiation wavelength. As given by eqn.(7), the polymerisation rate v_p is dependent on the square root of $(1-\exp(-2.303\varepsilon[PI]l))$. Using our model system TPGDA/IC 369, this relationship was experimentally verified up to a photoinitiator concentration of 1 wt.-%. At absorbances > 0.1 the approximation used to derive eqn.(7) is no longer valid.

Figure 5 shows a plot of the maximum polymerisation rate v_{pmax} vs. $(1-\exp(-2.303\varepsilon[PI]l))^{1/2}$.

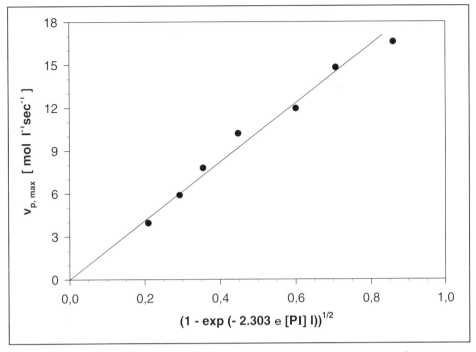

Figure 5 Polymerisation Rate v_{pmax} vs. $(1-\exp(-2.303\varepsilon[PI]l))^{1/2}$

As expected, at a photoinitiator concentration of 3 wt.-% the maximum polymerisation rate is only slightly larger than that measured for 1 wt.-% of photoinitator.

(v) Quantum yield of photoinduced polymerisation

The quantum yield of photoinduced polymerisation reflects the photochemical efficiency of the monomer conversion and is defined as:

$$\Phi_d = \frac{\text{Number of monomer molecules consumed}}{\text{Number of photons absorbed}} \quad (9)$$

It can be calculated from the ratio of the polymerisation rate to the number of photons I_a absorbed per second in the sample volume:

$$\Phi_d = v_p/I_a = \frac{v_p \ (\text{mol l}^{-1} \text{ s}^{-1}) \ l \ (\text{cm})}{I_0 \ (\text{mol s}^{-1} \text{ cm}^{-2}) \ 10^3 \ (1 - \exp(-2.303 \ e[PI]l))} \quad (10)$$

where l is the sample thickness, I_0 the incident photon flux and $\varepsilon[PI]l$ the optical absorbance of the sample.

For our model system TPGDA/0.3 wt.-% of IC 369, a maximum polymerisation quantum yield of $\Phi_d = 175$ was measured at a 313 nm irradiance of 48 mWcm^{-2}. Using a radical yield $\Phi_i = 0.22$ for IC 369 [9] about 800 monomer molecules were consumed per photoinitiator radical.

(vi) Effect of oxygen on photoinitiated polymerisation - the induction period

As illustrated in Figure 2, there is a time delay between the opening of the shutter at exactly 1 s after the start of the experiment and the onset of monomer conversion. This time delay, which is also called "induction period", is caused by the reaction of oxygen with the triplet state of the photoinitiator and/or the radicals formed after photoinitator triplet decomposition.

In the following example it is assumed that a typical scission-type photoinitiator PI absorbs a photon and transforms into an excited triplet state PI^T. In the presence of oxygen, triplet quenching by oxygen competes with decomposition into the radicals R_1 and R_2. Oxygen diffusion into or out of the sample during exposure is assumed to be negligible. Air saturation of acrylates lead to an oxygen saturation concentration of about 2×10^{-3} mol l^{-1}. This value is regarded as the initial oxygen concentration. Oxygen depletion can be induced by triplet formation and reaction with photoinitiator radicals. The following simplified reaction scheme will be used to estimate the duration of the induction period for our model system TPGDA/ 0.3 wt.-% of IC369:

Triplet formation	$\rightarrow$ PI^T	$v_T = \Phi_T I_a$
Triplet quenching	PI^T + O$_2$ $\rightarrow$ products	$k_q[O_2)]_0 = 5 \times 10^9 \times 2 \times 10^{-3} = 10^7$ s^{-1}
Triplet decomposition	PI^T $\rightarrow$ R$_1$ + R$_2$	$k_d = 0.59 \times 10^5$ s^{-1}
Radical quenching	R$_{1/2}$ + O$_2$ $\rightarrow$ R$_{1/2}$OO$^\bullet$	$k_{rq} = 1 \times 10^9 \times 2 \times 10^{-3} = 2 \times 10^6$ s^{-1}
Initiation	R$_{1/2}$ + M $\rightarrow$ P$^\circ_{1/2}$	$k_i = 10^5 \times 6.6 = 6.6 \times 10^5$ s^{-1}
Radical quenching	P$^\circ_{1/2}$ + O$_2$ $\rightarrow$ P$_{1/2}$OO$^\bullet$	$k_{pq} = 10^7 \times 2 \times 10^{-3} = 2 \times 10^4$ s^{-1}

Figure 6 Effect of Oxygen on the Initiation of Photoinduced Radical Polymerisation

On the right hand side of Figure 6, rates (in s^{-1}) for the different reactions were estimated. The triplet quenching rate $k_q[O_2]_0$ is about one order in magnitude larger than all other rates of competing reactions. For our model system TPGDA/IC 369, triplet quenching seems to be the dominating pathway demonstrating how oxygen interferes with radical formation. Immediately after the start of light exposure, the triplets generated by a rate

$v_T = \Phi_T I_a$ exclusively react with oxygen. Oxygen depletion takes place in the sample until a residual oxygen concentration is reached which neither affects the competing photoinitiator decomposition nor the initiation of polymerisation.

Assuming the reaction mechanism illustrated in Figure 6, the maximum induction period Δt_i can be estimated by

$$\Delta t_i = [O_2]_0 / v_T = [O_2]_0 / \Phi_T I_a . \tag{11}$$

$$\Delta t_i = [O_2]_0 / \Phi_T I_o (1 - \exp(-2.303\varepsilon[PI]l)). \tag{12}$$

It is of practical importance in UV curing that the induction period decreases at increasing irradiance and absorption.

For our model system TPGDA/ 0.3 wt.-% of IC369 a maximum induction period of 250 ms was estimated at a 313nm irradiance of 48 mWcm^{-2}. This value compares well with the observed induction period (see Figure 2).

It should be pointed out, however, that some other photoinitiators such as α-hydroxy alkylphenones and acylphosphine oxides have triplet lifetimes of a few nanoseconds. In this case, triplet quenching by oxygen plays a minor role and radical quenching gains in importance (see Figure 6).

The simple estimate of induction period duration as given by eqns. (11) and (12) is no longer valid, if the sample is not covered by a quartz plate and oxygen diffusion can take place.

If curing is done under air, oxygen diffusion into the sample prevents its fast depletion and much higher triplet formation rates are needed to reduce the induction period.

On the other hand, if nitrogen inerting of the sample is applied, oxygen diffuses out of the sample, and the oxygen concentration present during exposure is lower than the saturation concentration. For a more quantitative discussion of the oxygen effect, diffusion profiles were calculated as a function of the diffusion coefficient and sample thickness. The results are presented below and discussed in the context of physical factors affecting cure speed.

(vii) Photochemical dark reaction (postcure)

After light exposure has finished, photoinitiated radical polymerisation continues to proceed in the timescale of seconds, minutes and sometimes even hours. Figure 7 shows, for example, the double bond conversion profile recorded during and after 313 nm irradiation of our model system TPGDA/1 wt.-% of IC 369. An irradiance of 40 mWcm^{-2} was measured in the ATR crystal plane and exposure was finished after 100 ms. Under these conditions, 30% double bond conversion was obtained immediately after exposure. However, within the following 11 s the conversion reached 63 %.

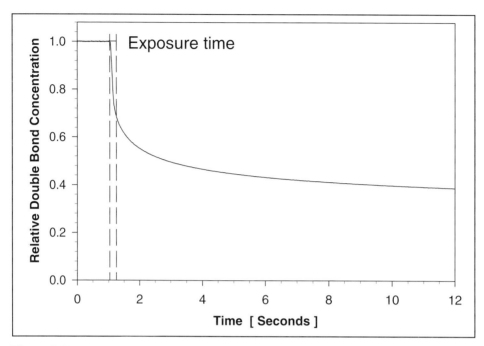

Figure 7 Double Bond Conversion Recorded During and After 313 nm Exposure of TPGDA/1 wt.-% of IC 369. Exposure Time 100 ms, Irradiance 40 mWcm^{-2}

This pronounced postcuring process continued for more than one hour. In the timescale of minutes postcuring could be followed by Electron Paramagnetic Resonance (EPR). In the experiment, polymerisation was induced by irradiating a degassed acrylate sample with a certain number of 308 nm laser pulses. Irradiation was stopped after the sample became solid, and noise-free radical spectra could be recorded. After exposure, the spectra were followed over more than one hour until the signals disappeared into noise. Figure 8 shows the double integral of the EPR spectrum (as a measure of radical concentration) recorded over time.

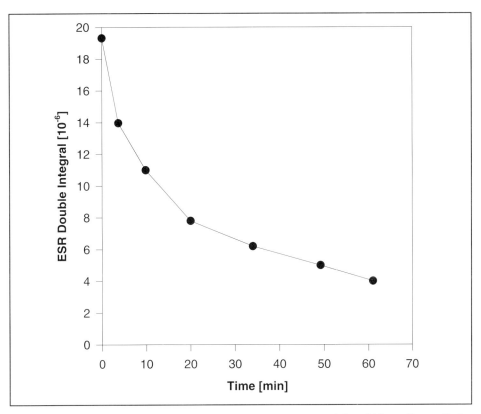

Figure 8 Decay of the Radical Concentration Recorded After 308 nm Laser Pulse Irradiation of an UV Printing Ink

From the EPR spectra two types of acrylate radicals were identified: that of the growing chain

and a so-called mid-chain radical.

Propagation and termination which can even take place in highly viscous and solid networks cause postcuring. In order to add to double bonds or to recombine, radical sites which are embedded in relatively rigid polymer structures have to migrate by non-diffusional displacement. Sometimes segmental motion of polymer chains leads to an approach of radical sites to double bonds or other radical sites. In the presence of residual oxygen, the displacement of the radical sites can be peroxy radical driven.

Figure 9 schematically illustrates a possible mechanism of radical site migration and postcuring by chain propagation and cross-linking (termination).

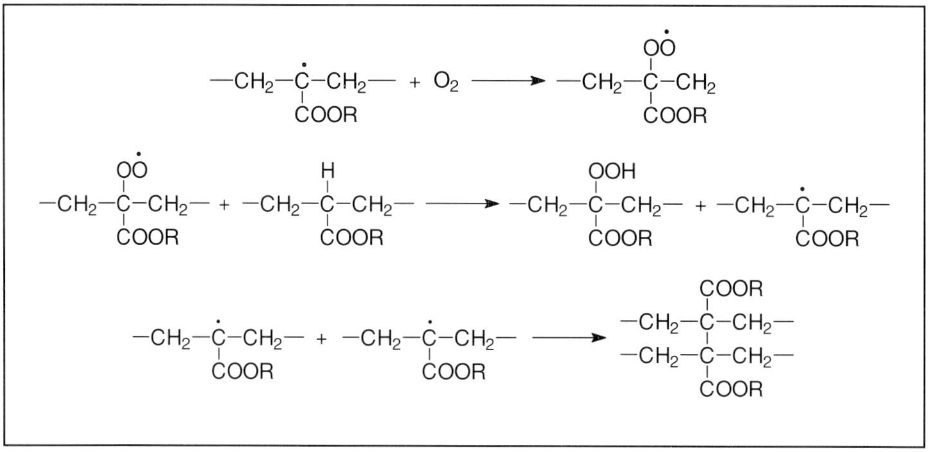

Figure 9 Possible Mechanism of Postcuring in Acrylates

(viii) Experimental determination of propagation and termination rate constants

Monomer conversion vs. time profiles as recorded during and after exposure, for example, by real-time FT IR spectroscopy, contain the information needed to deduce the rate constant of chain propagation k_p as well as that of termination k_t. Under stationary irradiation conditions, the ratio $k_p (k_t)^{-1/2}$ can be obtained from the slope α of the plot of the polymerisation rate v_p vs. the square root of Ia (see Figure 4). Monomer conversion vs. time as measured after light has been switched off (see Figure 7) can be used to calculate the ratio k_t/k_p [7,10].

After exposure the concentration of polymer radicals $P°_n$ will decrease by termination:

$$- d[P°_n]/dt = k_t [P°_n]^2 \quad \text{or} \quad 1/[P°_n] = k_t t + 1/[P°_n]_o , \qquad (13)$$

where $[P°_n]_o$ is the radical concentration at the end of irradiation.

$[P°_n]_o$ can be expressed as a function of the propagation rate $(v_p)_0$ and the monomer concentration $[M]_0$ at the end of exposure:

$$(v_p)_0 = k_p [P°_n]_o [M]_0 . \qquad (14)$$

The polymer radical concentration $[P°_n]$ reached at a given time t after exposure can be obtained from

$$[P°_n] = (v_p)_{0+t} / k_p [M]_{0+t} \qquad (15)$$

By replacing $[P°_n]$ and $[P°_n]_o$ in eqn.(13), the following rate equation for postcuring is obtained:

$$[M]_{0+t}/ (v_p)_{0+t} = (k_t/k_p) t + [M]_0/(v_p)_0. \qquad (16)$$

Propagation rates and monomer concentrations can be determined from the conversion vs. time profiles taken by real-time FT IR. Individual values of k_t and k_p can be calculated from $k_p(k_t)^{-1/2}$ and (k_t/k_p) obtained during exposure and postcuring, respectively.

(ix) Effect of temperature on the polymerisation rate and the degree of cure

During the UV or EB curing process, the sample temperature rises due to heat generated by the exothermal polymerisation reaction. In the case of UV exposure heat flow rate (dH/dt) and temperature increase ΔT can be monitored by calorimetric methods such as Photo-DSC [1,3]. However, the poor time resolution of DSC and the large sample volume needed create unfavourable conditions to record temperature profiles of very thin (thickness of a few µm) samples at intense UV irradiation. Additionally, due to technical reasons such as radiation shielding, monitoring of EB curing by DSC has been not tried. Therefore, the following simple technique was applied to directly measure temperature vs. time profiles in electron-initiated polymerisation [11,12].

A thin acrylate sample was placed on a polyimide foil and cured by irradiation with 180 keV electrons. The heat flux generated by the evolving temperature rise in the irradiated sample was monitored by a calibrated fast (response time < 100 ms) thermosensitive detector and recorded using a transient recorder. The detector was placed on the conveyor of the irradiation unit and fed through the electron "curtain" of a low energy electron accelerator (see Figure 10a). Simultaneously, the relative electron current was monitored as a function of time by a Faraday cup placed in line with the thermodetector.

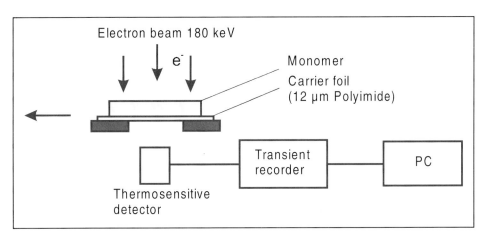

Figure 10a Experimental Set-up for Monitoring the Temperature Rise in Electron-Irradiated Acrylate Samples

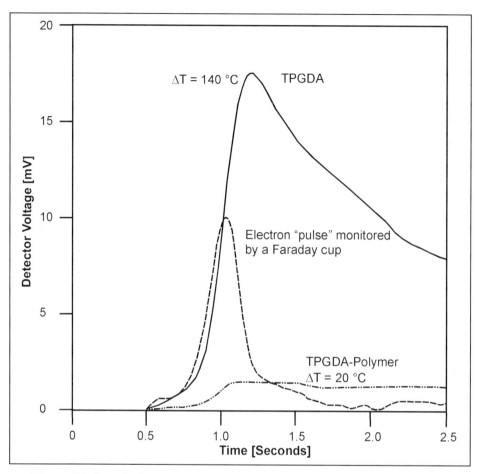

Figure 10b Sample Temperature as Function of Time Measured for Electron-Irradiated TPGDA. Electron-Irradiation Profile Monitored by a FARADAY Cup.

During EB curing of a of 120 µm thick TPGDA layer a maximum temperature rise ΔT of about 120°C is obtained (see Figure 10b). The temperature rise measured in a second run for the electron absorber consisting now of the acrylate polymer and the foil, is due to electron absorption and can be related to the absorbed dose.

ΔT decreases as heat dissipation becomes important. Because thermoisolation of the sample is not ideal, the adiabatic temperature rise is expected to be always higher than the observed non-adiabatic one.

Conversion vs. time profiles as recorded by real-time FTIR can also be used to determine the adiabatic temperature rise during photoinitiated polymerisation.

The number of monomer moles per litre reacted at time t, $[M_0]-[M_t]$, can be easily calculated from the experimental conversion profiles (see Figure 2). For acrylate monomers the standard heat of polymerisation ΔH_0 is known to be in the range of 78 to 86 kJ per mole of double bonds. The heat of polymerisation at time t, ΔH_t, can simply be calculated by:

$$\Delta H_t = ([M_0]-[M_t])\Delta H_0, \qquad (17)$$

The (adiabatic) temperature rise ΔT in the sample is obtained from:

$$\Delta T = ([M_0]-[M_t])\Delta H_0 / c_p, \qquad (18)$$

where c_p the specific heat capacity.
Using $\Delta H_0 = 78$ kJmol^{-1}, $c_p = 1.5$ kJ kg^{-1} degree^{-1} and 90% double bond conversion for the model system TPGDA/1 wt.-% of IC 369 an adiabatic temperature rise $\Delta T = 294°C$ is estimated.

It is not an easy endeavour to calculate the non-adiabatic temperature vs. time profile of a curing sample. The heat transfer from the sample into the substrate is extremely different. Under the experimental conditions of real-time FTIR as illustrated in Figure 1 (sample thickness 4µm, substrate consisting of a highly heat conducting metal and diamond) the internal temperature rise can be neglected. For metal substrates this seems to be the case in general.

At a higher sample thickness, for more reactive monomers and good thermoinsulation, the real temperature increase in a sample can even reach 80-100°C. If the heat evolving cannot be dissipated, for example, if coated flexible substrates are rewound immediately after curing, temperature sensitive substrates can be damaged.

The "internal" temperature rise occurring during polymerisation must be kept small, when the effect of „external" temperature on photoinitiated polymerisation is studied. In this context, the temperature dependence of the polymerisation rate v_p is the most significant parameter. It is expressed by the total activation energy of the polymerisation process.

As mentioned above, the rate of photoinitiated polymerisation is given by:

$$v_p = k_p (k_t)^{-1/2} [M] (\Phi_i I_a)^{1/2}. \qquad (4)$$

Applying the Arrhenius equation

$$K = A \exp(-E/RT), \qquad (19)$$

where $K = k_p (k_t)^{-1/2}$, A is the preexponential factor taking into account the frequency of molecular collisions and RT is the temperature energy, we obtain the following expression for the total activation energy E of polymerisation:

$$E = E_p - 1/2 E_t, \qquad (20)$$

where E_p and E_t are the activation energies of propagation and termination, respectively.

Plotting $\ln v_{p,\,max}$ vs. $1/T$, the activation energy E can determined from the slope of the Arrhenius curve.

To elucidate the temperature dependence of UV curing, the model formulation TPGDA/0.1 wt.-% of IC 369 was studied at various temperatures in the range of 25 up to 100 °C. A layer of 4 µm was exposed for 10 seconds to an irradiance of 48 mWcm^{-2} at 313 nm. Simultaneously, infrared spectra were recorded with a time resolution of 11 milliseconds.

The conversion of the acrylic double bonds is obtained from the decay of the absorption band of the $=CH_2$ twisting vibration at 810 cm^{-1}. The kinetic profiles for the photopolymerisation of TPGDA at various temperatures are plotted in Fig. 11 [13].

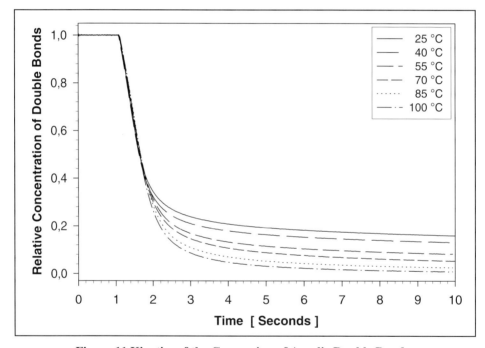

Figure 11 Kinetics of the Conversion of Acrylic Double Bonds in TPGDA/ 0.1 wt.-% IC 369 During Irradiation at Various Temperatures with 48 mWcm^{-2} at 313 nm

The shutter was opened exactly 1 second after the start of spectra recording.

The slopes of the conversion curves recorded at different temperatures do not show any observable change.

However, the degree of cure is considerably improved by an increase in temperature. A rise of the conversion of double bonds from 82% at 25°C to 92% at 55°C is achieved.

An explanation for this remarkable effect could be that a temperature rise induces improved segmental motion of the polymer chains and makes more residual unsaturation sites accessible to polymerisation.

From the kinetic curves in Fig. 11, the polymerisation rate v_p was calculated. Results are shown in Fig. 12. The kinetic behaviour of the model system was measured at least 5 times for each temperature. Thus, curves in Fig. 12 are averaged polymerisation profiles.

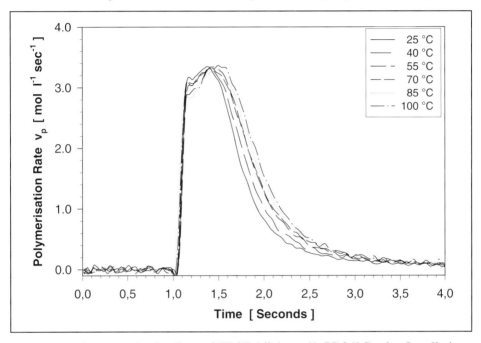

Figure 12 Polymerisation Rate of TPGDA/0.1 wt.-% IC 369 During Irradiation with 313 nm at Various Temperatures

It can be clearly seen from Figure 12 that temperature has practically no effect on the polymerisation rate of TPGDA in the temperature range studied.

A plot of ln $v_{p.\ max}$ vs. 1/T reveals that behaviour. Figure 13 shows that, within the experimental error limit, the activation energy of the polymerisation rate is zero.

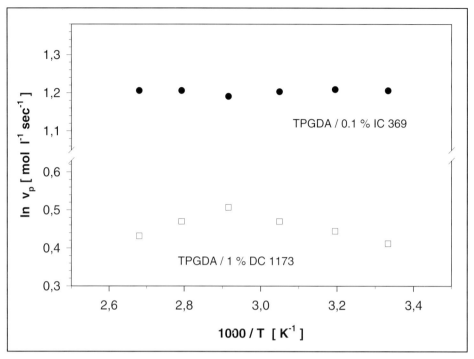

Figure 13 Arrhenius Plot of the Maximum Polymerisation Rate of TPGDA/ 0.1 wt.-% IC 369, Irradiation Conditions as given in Figure 11

(x) **Effect of temperature on the induction period**

Again real-time FTIR with ATR detection proved to be an useful technique to study the effect of temperature on the duration of the induction period. For the model formulation TPGDA/1 wt.-% of DC 1173 the duration of the induction period decreased at rising temperature. This effect is shown in Figure 14. A similar behaviour was also observed for other acrylate/photoinitiator combinations. Due to the complex mechanism of oxygen interference with photoinitator decomposition and the initiation of polymerisation, an explanation of the observed temperature effect is difficult. However, one should keep in mind that the solubility of oxygen in acrylates drastically reduces at higher temperatures. The triplet quenching rate decreases and competing photoinitiator decomposition gains in importance.

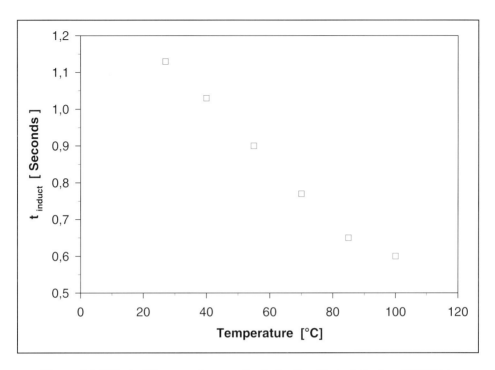

Figure 14 Effect of Temperature on the Induction Period, System TPGDA/ 1 wt.-% of IC 1173, Irradiation with 313 nm , Irradiance 48 mWcm^{-2}

3.   Cure speed

In the real world of curing technology, the pragmatic concept that,

> *the cured product has to meet the needs of its function properties to be commercial,*

is not only used as a definition of (a sufficient degree of) cure, but also the term *cure speed* is frequently applied to compare the reactivity of different coatings, paints or inks towards UV irradiation. In a typical UV curing experiment, the coating to be cured by a certain speed is passed through the irradiation zone of a lamp/reflector combination.

The maximum (conveyor, belt, production etc.) speed at which the *cured product still meets the needs of its function properties to be commercial* is called *cure speed*.

The cure speed defined in such a way can only be used as a relative measure of the reactivity of different UV curing formulations, if the following physical and technological curing conditions are kept constant:

- spectral distribution of the lamp,
- irradiance and irradiance distribution in the curing plane,

- thickness of the coating,
- shape of the surface to be cured,
- oxygen concentration (air or inert),
- coating temperature (within a certain range)

and if reliable test methods are available to distinguish whether or not the cured product meets the function properties desired.

Of course, for a given formulation the cure speed can be studied as a function of all the parameters mentioned above. In that sense, consistent experimental data can be obtained.

The weakness of the cure speed concept, which is not based on a clean kinetic approach as if was for the polymerisation rate, is illustrated in Figure 15.

As shown in the upper part of Figure 15, a certain irradiance distribution is seen by a coating increment dF passing the exposure zone Δx with a uniform speed v_s. Within a total irradiation time $\Delta t_{exp} = \Delta x/v_s$, the induction period Δt_i has to be overcome, and after the conversion time Δt_{con} the monomer conversion must reach a distinct level. The cure speed is then given as:

$$v_s = \Delta x/(\Delta t_i + \Delta t_{con}).$$

After passing the exposure zone postcuring takes place in most cases. Therefore, it is not well defined at what time curing can be regarded as giving the desired product function properties.

For example, immediately after UV curing printing inks sometimes are not "through-cured" and can be wiped out by thumb twists. In most cases, however, through-curing is obtained by postcuring some minutes or hours later.

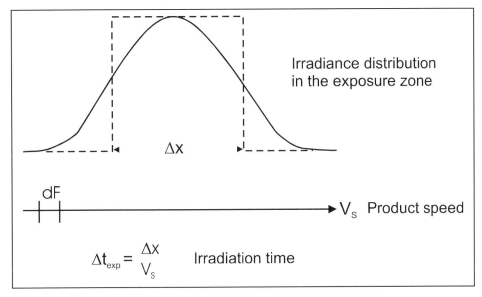

Figure 15 Curing Process and Cure Speed

The maximum obtainable cure speed is of great practical significance. In many industrial curing applications the cure speed is an important factor controlling the economy of the process. Therefore, the effect of chemical and physical factors on cure speed is outlined below in more detail.

4. Chemical and physical factors affecting the cure speed

(i) Effect of chemical factors

(a) Functionality

Due to the complexity of the cure speed concept, all chemical factors are of importance in this context as they affect the induction period, the polymerisation rate and postcuring.

It is well known that the functionality of acrylates has a strong influence on both the polymerisation rate and the residual monomer content. With increasing acrylate functionality, the conversion time decreases but the content of residual unsaturations rises. This behaviour is illustrated in Figure 16.

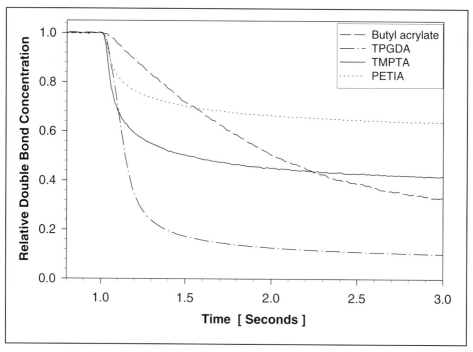

Figure 16 Effect of the Acrylate Functionality on Conversion Time and Residual Monomer Concentration. 1 wt.-% Photoinitator added to the Acrylates, Irradiance at 313 nm 74 mWcm^{-2}

As the functionality increases, the higher initial concentration of acrylate groups leads to initially faster conversion but the higher viscosity of the resin, with the resulting gel-effect and the higher cross-link density, set a limit to the extent of conversion.

(b) Photoinitiator

The nature and concentration of the photoinitiator used in the formulation also affect the cure speed. Under constant physical exposure conditions both induction period and polymerisation rate are influenced (see Figure 17).

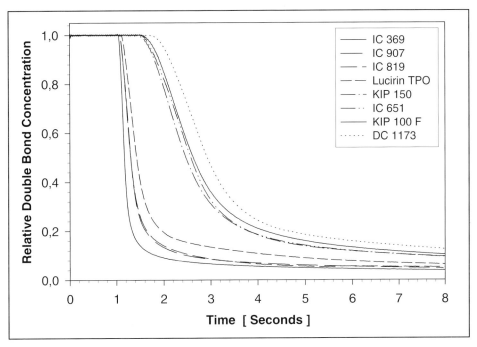

Figure 17 Double Bond Conversion in TPGDA Measured for Different Photoinitiators, Photoinitiator Concentration 1 wt.-%, Irradiance 40 mWcm^{-2} at 313 nm

To obtain a high polymerisation rate and a short induction period, the photoinitiator absorption spectrum and the emission spectrum of the curing light source should overlap as much as possible. Because in 95% of all UV-curing applications medium pressure mercury lamps are used, the photoinitiator absorption spectra were tuned to absorb strongly at intense mercury emission lines. As an example, Figure 18a shows this overlap for IC 369. Fine-tuning of the photoinitiator absorption to monochromatic excimer radiation remains still to be done. In the case of 308 nm emission, however, photoinitiators which were originally tuned to mercury emission lines between 300 and 313 nm can be used favourably. Figure 18b shows an example using the photoinitiators IC 369, 907 and 819. Their high extinction coefficients at 308 nm are a prerequisite for fast conversion in monochromatic UV-curing.

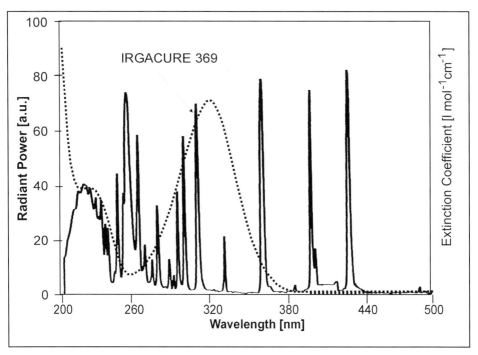

Figure 18 a) Spectral Overlap of Photoinitiator (IC 369) Absorption and Emission from a Medium Pressure Mercury Lamp

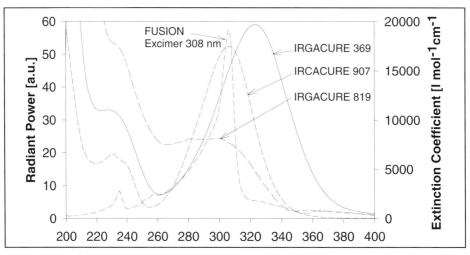

Figure 18 b) Spectral Overlap of Photoinitiator (IC 369, 907 and 819) Absorption and Emission from a 308 nm Excimer Lamp (FUSION UV Systems)

(c) Pigmentation

In pigmented systems pigment and photoinitiator absorptions often superimpose. The pigment can absorb a considerable part of the incoming photons flux. As a result, at constant exposure conditions the induction period slightly increases and the polymerisation rate drops. Figure 19 shows the effect of the pigment content on double bond conversion time for an ink formulation.

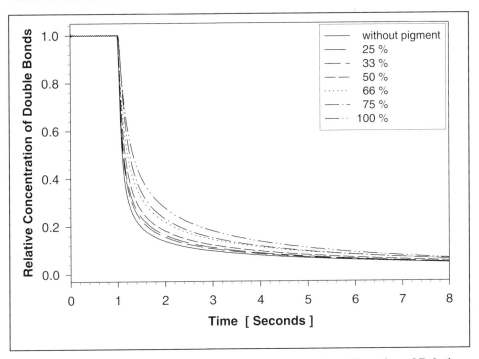

Figure 19 Double Bond Conversion in a Printing Ink as Function of Relative Pigment Concentration. Irradiance at 313 nm 48 mWcm^{-2}

For some pigmented systems it is possible to improve the cure speed by using photoinitiators with optical absorption outside the pigment absorption and UV sources with increased spectral output at higher wavelengths.

(ii) Effect of physical factors

(a) Irradiance in the curing plane

It is obvious from equation (4) and (11):

$$v_p = k_p (k_t)^{-1/2} [M] (\Phi_i I_a)^{1/2}, \qquad (4)$$

$$\Delta t_i = [O_2)]_0 / v_T = [O_2)]_0 / \Phi_T I_a \qquad (11)$$

that the polymerisation rate v_p as well as the induction period Δt_i are dependent on I_a, the number of photons (given in moll^{-1}) absorbed per second within the sample. I_a is directly proportional to I_o, the number of photons impinging per cm^2 sample surface and second. I_a further depends on the photoinitiator absorbance.

The irradiance measured in the curing plane is proportional to I_o, i.e., the polymerisation rate should increase according to the square root of irradiance, whereas the induction period should decrease linearly at growing irradiance. The polymerisation rate determines a conversion time Δt_{con} at which a certain degree of cure is obtained. On the other hand, the minimum exposure time Δt_{exp} to reach the desired conversion is the sum of induction and conversion time. At increasing irradiance the induction time Δt_i decreases linearly but the conversion time decreases according to the square root:

$$\Delta t_{exp} = \Delta t_i + \Delta t_{con} = [O_2]_0/\Phi_T I_a + \beta/(\Phi_i I_a)^{1/2}, \qquad (12)$$

where β is a proportionality factor.

As a result of eqn.(12) it is expected that the cure speed $v_s = \Delta x/\Delta t_{exp}$ rises less than linearly with growing irradiance, for example, if irradiance is increased by a factor of two the resulting exposure time is longer than half the initial one.

Eqn.(12) was derived under the assumption that oxygen diffusion does not take place during exposure. Application of eqn.(4) further implies that:

- monochromatic light is used, which is absorbed exclusively by the photoinitiator
- the absorption is small and homogeneous in the volume irradiated and
- the extent of conversion is low.

As mentioned above, this is an idealised case. There is experimental evidence that UV curable coatings may respond to changes in UV irradiance. By increasing the irradiance at constant dose, improved chemical resistance of the coating or higher cure speeds have been reported [14,15].

(b) Oxygen inhibition

In UV curing, reactions of photoinitiator triplet, initiating and propagating radicals with oxygen lead to an induction period before monomer conversion can take place. To induce curing the exposure time has to be longer than the induction period. Thus, the oxygen content of the sample can strongly affect the cure speed. At low irradiance and between the sample and the surrounding air, curing can even be prevented by oxygen. The cured coating or ink remains tacky, an effect which is frequently called oxygen inhibition. This effect is often restricted to thin surface layers and it is less pronounced or even disappears at increasing depth. In UV curing under air, oxygen inhibition is a result of the parallel action of oxygen depletion by irradiation and oxygen diffusion into the sample. The driving force of diffusion is the oxygen concentration difference between surrounding air and sample. Concentration and time profiles of the diffusing oxygen can be calculated by solving the (one-dimensional) diffusion equation:

$$\partial c/\partial t = D\partial c/\partial x, \qquad (13)$$

where c is the oxygen concentration and D the diffusion coefficient of oxygen within the matrix. The diffusion coefficient depends on the temperature and viscosity of the formulation. Because the matrix viscosity changes over more than three orders in magnitude during curing, to estimate diffusion profiles a range of diffusion coefficients has to be used. Using the diffusion coefficients given in Table 2 the diffusion profiles shown in Figure 20 were calculated for different times.

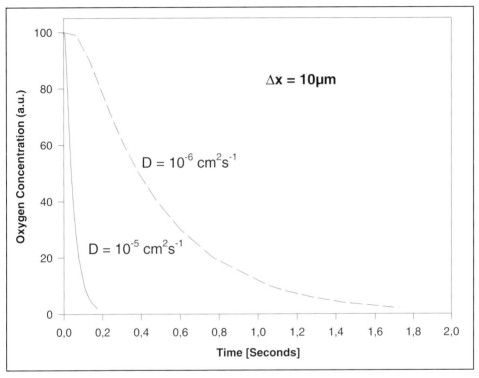

Figure 20 Diffusion Profiles Calculated for Oxygen in a 10 μm Layer. Diffusion Coefficent D indicated.

To estimate the effect of oxygen diffusion under real curing conditions, an even simpler model can be applied. The mean distance Δr that diffusing oxygen can travel during a time Δt can be expressed as:

$$\Delta r = (2Dt)^{1/2} . \tag{14}$$

Table 2 contains the Δr values calculated for different diffusion coefficients after $\Delta t = 100$ ms.

This time is selected because it is in the range of typical exposure times and induction periods.

Table 2 Diffusion Coefficients for Oxygen and Mean Diffusion Distances after 100 ms

	Liquid		Highly viscous or solid	
Diffusion coefficient of oxygen [$cm^2 s^{-1}$]	10^{-5}	10^{-6}	10^{-7}	10^{-8}
Mean diffusion distance [µm]	14.1	4.5	1.4	0.45

It is obvious from Table 2 that in a low-viscosity liquid, diffusion is fast and can easily compete with oxygen depletion up to a depth of about 10 µm. In practice, this effect is noticed as tacky surface. The diffusion depth is strongly reduced at a higher viscosity but is still of importance. This is for example the case for offset printing inks, where the ink thickness is in the range of 1µm.

If UV curing of coatings or inks occurs in contact with air, a considerable amount of photons is wasted in order to reduce the oxygen concentration. Additionally, volatiles are generated via peroxy radicals, which can contribute to an undesired odour of the cured products.

(c) Nitrogen inerting

To avoid oxygen inhibition, nitrogen inerting (sometimes also called nitrogen blanketing) is the preferred technique. In nitrogen inerting the oxygen surrounding the coating and adhering on the coating surface is removed by a laminar nitrogen flow. Using this technique, residual oxygen concentrations of ppm can be obtained in the inerting gas. Inerting is usually done in a closed chamber, a few hundred milliseconds before UV curing takes place. Effective oxygen diffusion from coating to the gas phase results in a considerable decrease of the dissolved oxygen. The induction period decreases and the coating undergoes faster curing.

Even if UV irradiance is high enough to enable UV curing under air, nitrogen inerting leads to important technological benefits:
- reduced energy consumption,
- reduced heat transfer to the substrate,
- no ozone production,
- reduced smell of the cured product,
- reduction of photoinitiator concentration possible,
- reduced amount of extractables from the cured product.

For most UV curing applications mature technical solutions for nitrogen inerting are available (see Chapter VII). In particular, for curing coatings and inks on paper and film the nitrogen consumption can be kept on a level at which the costs of nitrogen inerting are compensated by savings in power costs. These are applications where high speed and high quality curing demands inerting.

(d) Temperature

The temperature dependence of the cure speed can be described by the combined action of three effects, which were already discussed above:
- the decrease of the induction period at a rising temperature,
- the small or even negligible effect of temperature on the polymerisation rate and
- the decrease of the residual unsaturation at increasing temperature.

The most pronounced temperature effect is that on the duration of the induction period. If the induction period is comparable to the monomer conversion time, the temperature of the coating markedly affects the cure speed. In inerted systems the induction time is, in general, small in comparison with conversion time. Here the third point comes into play: the higher degree of cure, which can be observed at increasing temperature, means that the conversion desired for the *product to be commercial* can be reached at a shorter conversion time.

(e) Multiple exposure

In many UV curing applications the desired production speed cannot be reached by using only a single lamp. In these cases the exposure time is too short to generate sufficient conversion. Therefore, it is common to use multilamp irradiation units. Figure 21 shows the experimental simulation of the curing behaviour of a coating which passes four lamps. Again the model system TPGDA/1 wt.-% of IC 369 was chosen. In one-second intervals, exposure pulses of 120 ms duration and equal intensity were repeated several times. Monomer conversion was followed by real-time FTIR. During the dark period between the light pulses pronounced postcuring takes place. After two exposures already (240 ms total irradiation time), a degree of cure of about 80% is obtained. A single exposure of 200 ms would produce a similar degree of cure.

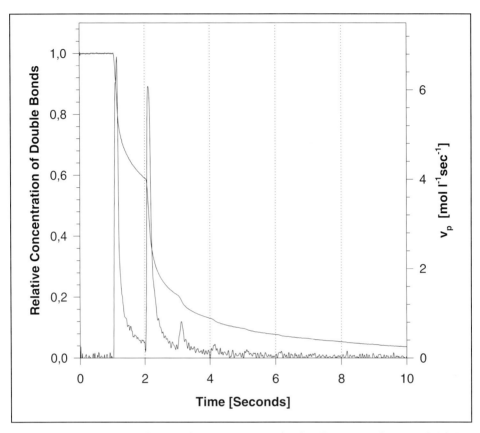

Figure 21 Monomer Conversion and Polymerisation Rate vs. Time at Multiple Pulse Exposure: Exposure Time 120 ms, Time between Two Pulses 1 s. System TPGDA/1wt.-% IC369, Irradiance at 313 nm 40 mWcm^{-2}

5. Evaluation of kinetic parameters in cationic curing
(i) Polymerisation rate

Figure 13 of Chapter I outlines the general mechanism of cationic curing. In order to evaluate kinetic parameters of cationic curing, it is useful to recall that the initiating species, e.g. H^+, is not a radical and is only consumed by anions or nucleophiles. Termination of the polymer cation takes place either by recombination with anions or by reaction with nucleophiles. Using the reaction scheme shown in Figure 22, the polymerisation rate of cationic polymerisation can be estimated.

Initiation	$H^+A^- + M \longrightarrow H-M^+ + A^-$	$v_i = \phi_i I_a$
Propagation	$H-M^+ + M \longrightarrow H-M_2^+$	$v_p = k_p[M^+_n][M]$
	$M_n^+ + M \longrightarrow MM_n^+$	
Chain Transfer	$M_n^+ + ROH \longrightarrow M_n-OR + H^+$	
Termination	$M_n^+ + A^- \longrightarrow M_nA$	$v_t = k_t[M_n^+][A^-]$
	$M_n^+ + N \longrightarrow$ products	$v_t = k_t[M_n^+][N]$

Figure 22 Reaction Steps and Rates in Cationic Polymerisation

Neglecting the chain transfer, initiation v_i, propagation v_p and termination rate v_t can be expressed by:

$$v_i = \Phi_i I_a, \quad v_p = k_p[M_n^+][M], \quad v_t = k_t[M_n^+][A^-]. \quad (15)$$

First we assume that nucleophiles are absent. Since the contents of ions of opposite signs is equal ($[M_n^+] = [A^-]$), their neutralisation is expressed by bimolecular termination. For this case follows:

$$v_p = k_p[M](\Phi_i I_a)^{1/2}/k_t^{1/2}, \quad (16)$$

where k_p and k_t are propagation and termination rate constants, [M] is the monomer concentration, I_a is the number of photons absorbed per second and Φ_i is the yield of initiating ions.

If electroneutral nucleophiles N, e.g. water, are present in the system with a concentration far exceeding that of the ions, the term $k_t[N]$ can be assumed as constant. In this case, termination is a monomolecular reaction, and the polymerisation rate

$$v_p = k_p[M](\Phi_i I_a)/k_t[N] \quad (17)$$

is proportional to I_a.

Depending on the termination mechanism, the exponent of $\Phi_i I_a$ should vary from 0.5 to 1.0 in the rate of photoinitiated cationic polymerisation.

To illustrate the theoretical considerations on UV curing kinetics given above by experimental examples, the following description uses experimental data which were obtained by real-time Fourier transform infrared spectroscopy (FTIR). Figure 1 of this chapter shows the scheme of the real-time FT IR apparatus used to obtain kinetic data on the cationic curing model systems consisiting of the cycloaliphatic diepoxide UVR- 6105 (75 wt.-%), the caprolactone polyol crosslinker Tone 0301 (25 wt.-%) and different amounts of the photoinitiator UVI-6990.

To analyse the conversion of the epoxy moiety as a function of time, the epoxy deformation vibration at 790 cm^{-1} was selected.

Figure 23 shows typical conversion vs. time profiles measured for the epoxy conversion of the model system containing 8 wt.-% photoinitiator and a cationic printing ink, respectively. The samples were continuously irradiated with monochromatic light (313 nm) at an irradiance of 70 and 48 mWcm^{-2}, respectively.

In Figure 24, the relative polymerisation rates derived from the conversion vs. time profiles of Figure 23b are given.

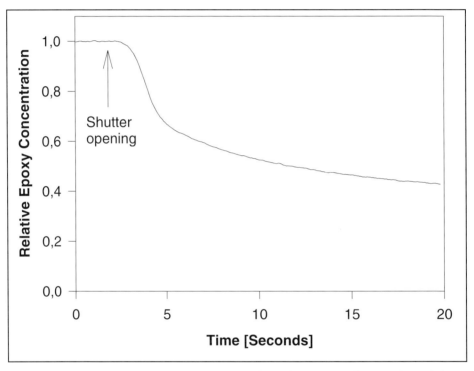

Figure 23a Kinetics of the Epoxy Conversion of the Model System Containing 8 wt.-% UVI 6990. Irradiation Conditions: Wavelength 313 nm, Irradiance 70 mWcm^{-2}, Shutter Delay Time 2 s.

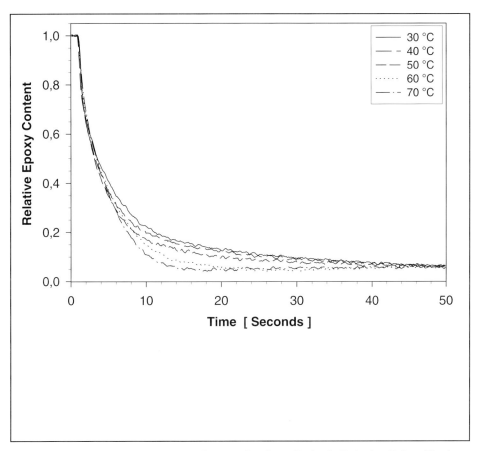

Figure 23 b Kinetics of the Epoxy Conversion in a Cationic Printing Ink at Various Temperatures. Irradiation Conditions: Wavelength 313 nm, Irradiance 48 mWcm^{-2}, Shutter Delay Time 1 s.

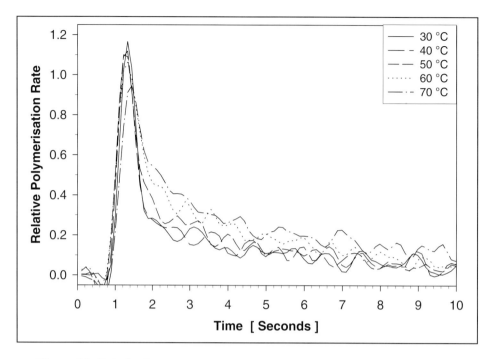

Figure 24 Relative Polymerisation Rates Derived from the Conversion vs. Time Profiles of Figure 23b

It is obvious from Figure 23 that:

- a small induction period is observed,
- compared to the radical UV curing model system (see Figure 2 in this Chapter) conversion is slow.

(ii) The polymerisation rate as a function of irradiance

As predicted by eqn. (16) and (17) a plot of the polymerisation rate v_p vs. I_0^n, the number of photons absorbed per second, should give a straight line with n between 0.5 and 1.

Using our model system UVR-6105/Tone 0301 containing 6 wt.-% of the photoinitiator UVI-6990, the maximum polymerisation rate v_p was measured as a function of irradiance I_0. As shown in Figure 25a, the conversion time is strongly dependent on the irradiance.

A plot of v_p vs. $I_0^{0.5}$ (see insert of Figure 25) shows nearly a straight line. At higher irradiances deviations from this functionality seem to occur.

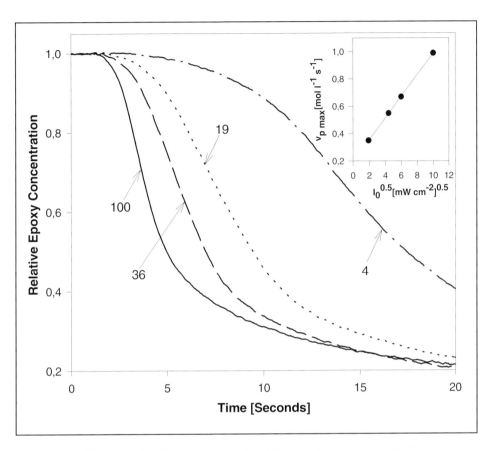

Figure 25a Relative Epoxy Conversion Measured at Different Irradiances. Irradiation Conditions: Irradiation with 313 nm, Irradiances given in mWcm^{-2}

(iii) The polymerisation rate as a function of photoinitiator concentration

Using Lambert-Beer's law the number of photons absorbed in the sample per second and unit area Ia can be calculated from the irradiance I_0, the photoinitiator concentration [PI], the sample thickness l and the photoinitiator extinction coefficient ε at irradiation wavelength:

$$I_a = I_0(1 - \exp(-2.303\varepsilon[PI]l)).$$

As calculated from eqns. (16) and (17) a plot of the polymerisation rate v_p vs. I_a^n should give a straight line for n between 0.5 and 1.

Using our cationic curing model system UVR-6105/Tone 0301 containing different photoinitiator concentrations this relationship was tested experimentally.

Figure 25b shows a plot of the maximum polymerisation rate $v_{p\,max}$ vs. (1-exp(-2.303 A)). The absorbance A was calculated from absorption spectra of the photoinitiator measured for a certain (solid) UVI-6990 photoinitiator concentration in g l^{-1} at an optical pathlength of 1 cm.

At absorbances up to about 0.15 (corresponding to 1-exp(-2.303A) = 0.29) an exponent of n = 1 is approached. This is expected when monomolecular termination of the cationic polymerisation takes place. As already mentioned in Chapter I, at higher absorbance values the derivation of the simple equations (16) and (17) is no longer valid. This could be the reason for the observed deviation of the polymerisation rate at photoinitiator concentrations higher than 6 wt.-%.

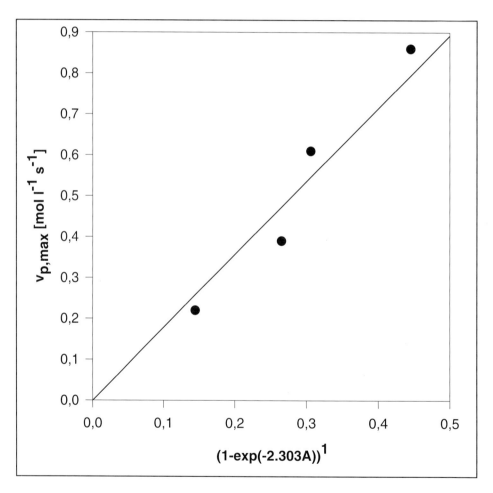

Figure 25b Polymerisation Rate $v_{p,\,max}$ vs. (1-exp(-2.303A))

(iv) Photochemical dark reaction (postcuring)

Pulsed UV irradiation of the cationic curing system and observation of the curing process using real-time FT IR is a excellent method to study the effect of the photo-induced dark reaction in cationic curing. In practical UV curing applications, pulsed exposure is the normal case. Figure 26 shows epoxide conversion vs. time profiles which were measured after exposure of the cationic curing model system with 313 nm light pulses of 400 ms and 1600 ms duration. For comparison, the conversion vs. time for continuous exposure is also given. The irradiance was kept constant at 75 mWcm^{-2}. The light shutter was opened after 2 s.

It is clearly seen that:
- an induction period of about 1 s takes place before the conversion begins,
- at 400ms exposure the dark reaction causes all the conversion,

The general conclusion can be drawn that in cationic UV curing the dark reaction plays a dominating role.

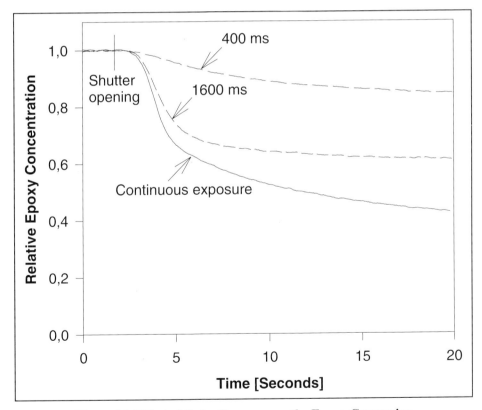

Figure 26 Effect of Pulse Exposure on the Epoxy Conversion

(v) Effect of temperature on the polymerisation rate

The results shown in Figures 23b and 24 indicate that the effect of temperature starts as soon as vitrification of the system sets in, i.e. when the glass transition temperature of the forming network exceeds the temperature of irradiation. Vitrification leads to a considerable decrease in the polymerisation rate. The mobility of the polymer chain segments and thus that of the remaining reactive groups can be increased by a temperature rise resulting in a progressive epoxide conversion. On the other hand, the reaction does not completely cease at lower irradiation temperatures (e.g. 30°C). In contrast to radically curable systems, the reaction continues slowly even after vitrification due to the longer life of the reactive species. The observed kinetic behaviour of the cationic model system can be explained by two phases of the curing process. At first, a rapid increase of epoxy conversion occurs. No influence of temperature is observed in this phase. After vitrification, the polymerisation is less affected by the continued irradiation with UV light, but by the temperature of the sample that determines the molecular mobility and consequently the reaction rate.

The polymerisation rate during the first 10 seconds of irradiation is plotted in Fig. 24. A relative polymerisation rate was calculated since the exact composition of the cationic printing ink is not known. The two phases of the polymerisation process can be clearly identified. During the initial phase of reaction, i.e. before the onset of sample vitrification, the temperature does not have any effect on the reaction rate. However, when the sample becomes glassy, the polymerisation rate shows a clear temperature influence. In the samples cured at 30 or 40°C, the rate rapidly slows down after the maximum, whereas the decay is much less pronounced at higher temperatures. Fig. 23b clearly shows that the reaction at 60 and 70°C is completed after about 15 seconds. In contrast, it very slowly continues at the other temperatures. Therefore, their rates of reaction should be different from zero.

6. Cure speed in cationic systems

The cure speed concept as illustrated for radical curing systems in Figure 15 has to be slightly modified for cationic systems.

As shown in the upper part of Figure 15, a certain irradiance distribution is seen by a coating increment dF passing the exposure zone Δx with a uniform speed v_s. Within a total exposure time $\Delta t_{exp} = \Delta x/v_s$ the monomer conversion must reach a certain level. That level is defined by the demand that:

the cured product meets the needs of its function properties to be commercial.

The cure speed is then given by vs = $\Delta x/\Delta t_{exp}$.

In comparison with radical curing, in cationic curing the photochemical dark reaction is even more pronounced. Sometimes it can be tolerated that the function properties of the cured product are reached after hours and days of postcuring.

7. Chemical and physical factors affecting the cure speed of cationic systems

(i) Effect of chemical factors

(a) Addition of alcohols and vinylethers to epoxies

As shown in Figure 22, hydroxyl group containing compounds such as alcohols can terminate the growing polymer cation by forming an ether linkage. The proton produced can react with a monomer unit, thereby initiating a new growing chain. This new chain is not bound to the network. Thus, higher polymerisation rates, i.e. cure speeds, can result from mobilisation by chain transfer as well as from a possible decrease in cross-link density [16].

However, chain transfer can lead to a reduction in mechanical strength and chemical resistance of the cured coating. This can be avoided by addition of multi-functional alcohols such as caprolactone polyols [17].

Another way to increase the cure speed of epoxies is the addition of vinyl ethers as reactive diluents or oligomers [18]. The vinyl ether component is responsible for the rapid change of the coating to a tack-free state but subsequent thermal treatment is often necessary to reach the desired degree of cure.

(b) Photoinitiators

Due to their good thermal stability and photosensitivity, triarylsulfonium salts have become the most frequently used photoinitiators for cationic UV curing (see Chapter I Figures 11 and 12). Solvated in propylene carbonate they are commercially available from a number of manufacturers. The absorption spectrum of triarylsulfonium cations overlaps very well with the emission spectrum of UV light sources such as medium pressure mercury lamps. Even around 308 nm, the emission maximum of XeCl excimer lamps, the extinction coefficient is high.

The inorganic counterions do not absorb light in the spectral region above 220 nm. They are not involved in photochemical or thermal reactions important for proton generation from sulphonium cations.

However, the nucleophilic character of the counterions has a considerable influence on cure speed and the degree of cure in cationic UV curing. The reactivity of the photoinitiators follows the decrease of nucleophility of the anions $SbF_6^- > AsF_6^- > PF_6^- > BF_4^-$. Recently, new borate anions $B^-(CF_5)_4$ have been associated with triarylsulphonium and iodonium cations. Using this anion, considerable improvement in reactivity was reported [19].

Another way to affect the generation of protonic acid from triarylsulphonium or iodonium salts is the electron transfer sensitation. In contrast to the direct excitation of the onium cations, the photosensitiser absorbs the light. The resulting excited sensitiser transfers an electron to the onium cation. Thus, a sensitiser radical cation and an onium free radical are formed. The onium radical rapidly decomposes generating an aryl radical and a stable product. Reaction of the aryl radical with the sensitiser cation finally leads to acid formation.

Aryliodonium cations are easily sensitised by aromatic hydrocarbons such as anthracene, pyrene or perylene, aromatic ketones and xanthones, whereas for sensitising triarylsulphonium cations aromatic hyrocarbons must be used.

In particular, sensitising is a suitable means to better match coating absorption and light source emission spectra.

(c) Water and nucleophiles

Water in the form of atmospheric moisture can adversely affect cationic curing. Already at a relative humidity of 50% residual surface tack can occur. Water as a nucleophile not only reacts with photogenerated protons but also with propagating carbonium ions. Proton transfer from the carbonium ion to the nucleophile is believed to be responsible for rapid termination, thus preventing complete cure.

Preheating the substrate or dehumidifying the cooling air of the UV lamp system can help to overcome cure problems. For cationic formulations additives and pigments with non-nucleophilic character should be selected.

(ii) Effect of physical factors

(a) Temperature

The benefit of heat applied during or after UV exposure to cationic UV curing systems can be twofold: firstly, the degree of cure is increased, secondly, when heated, water is taken up from the coating surface by the surrounding air.

In practical applications medium-wavelength infrared irradiation is the method of choice to finish dark curing within a short time and to produce coatings with a very low amount of uncured material.

(b) Oxygen effect and nitrogen inerting

Neither photogenerated acids as chain initiating species nor carbocations as chain propagating intermediates are sensitive to oxygen, i.e. cationic UV curing is not affected by oxygen. However, in practical curing applications it is sometimes observed that inerting of the irradiation zone leads to a remarkable increase (up to a factor of two) in cure speed. It is a side effect of inerting which is operative here: nitrogen gas as delivered from the evaporator is extremely dry. Using nitrogen within the exposure zone reduces the moisture content to a level practically not affecting the cure process.

8. Evaluation of kinetic parameters in electron beam curing
(i) Mechanism of electron beam curing

Electron initiated polymerisation can proceed by both radicals or ions. It depends on the nature of the monomer and of external conditions which mechanism is dominating. Electron irradiated acrylates show radical curing, whereas, e.g. for isobutene, ionic curing is observed. In the case of styrene, radical as well as ionic polymerisation can be induced by electron irradiation. The latter proceeds after rigorous drying even at low temperatures.

A general scheme for the electron induced formation of monomer (M) ions and radicals is given in Figure 27.

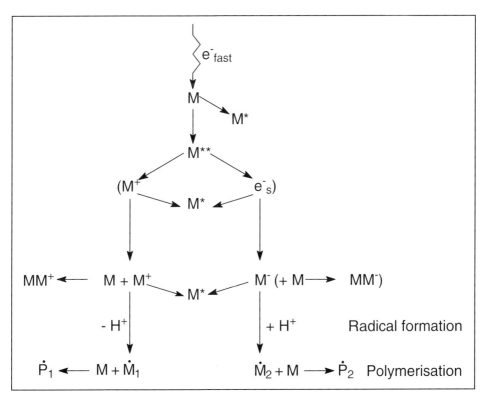

Figure 27 Electron Induced Ion and Radical Formation in Acrylate Monomers M

Coulomb interaction of fast electrons with monomer molecules leads to both ion pair formation (M^+ - e_s^-) and electronic excitation (M^*). Most of the primary ion pairs cannot escape their mutual Coulomb attraction and recombine by generating excited molecules. However, due to diffusional motion some ions become "free", i.e. they are not coupled to their primary partners any more. Monomers such as acrylates or styrene rapidly react with solvated electrons (e_s^-) forming anions. Both, cations and anions can start ionic polymerisation. Competing processes are ion recombination or radical formation. Cations often undergo deprotonation to generate neutral radicals. Anions can be protonated to become neutral radicals. Thus, as in the case of electron beam curing of acrylates, chain initiating radicals are usually produced via primary and secondary ions.

(ii) Polymerisation rate at radical initiation

To determine the polymerisation rate of the electron induced radical polymerisation the general scheme given in Figure 28 can be used.

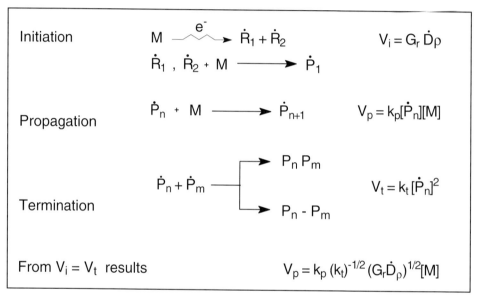

Figure 28 Steps in Electron Initiated Radical Polymerisation

In the mechanism it is only the initiation step which is different from photoinduced radical polymerisation. Electron irradiation plays the role of initiator. The rate of the absorbed dose $dD/dt = D°$ determines the initiation rate v_i:

$$v_i = G_r D° \rho, \tag{18}$$

where G_r is the radiation chemical yield of the initiating radicals, and ρ is the density of the monomer. The yield G_r is expressed in mol J^{-1}, the dose rate in Gy = J kg^{-1}s^{-1}, and ρ in kg l^{-1}.

Using the stationarity condition $v_i = v_t$, the polymerisation rate v_p is given by:

$$v_p = k_p (k_t)^{-1/2} (G_r D° \rho)^{1/2} [M]. \tag{19}$$

Propagation and termination of the electron initiated radical polymerisation proceeds just as for conventional radical polymerisation. The propagation rate constants are in the order of $k_p \approx 10^2\text{-}10^4$ l mol^{-1}s^{-1}, and that of $k_t \approx 10^5\text{-}10^7$ lmol^{-1}s^{-1}.

In practical UV curing applications ionic polymerisation is of minor importance. Vinyl ethers which in bulk polymerise by a cationic mechanism are frequently used as mixtures with acrylates, maleates or maleimides. In these systems copolymerisation occurs by a radical mechanism.

In ionic polymerisation termination can be either bimolecular, by ion recombination or monomolecular if electroneutral proton acceptors are present as impurities (e.g. water). Bimolecular termination leads to an exponent of 0.5 in equation (19) for the polymerisation rate constant, whereas in the extreme case of monomolecular termination an exponent of 1.0 is derived. In practice, with increasing monomer purity the exponent approaches 0.5. The high termination rate constants for ionic polymerisation given in Table 3 point to ion recombination as the dominating termination process.

To illustrate this behaviour, Table 3 summarises propagation and termination rate constants of radical and ionic polymerising monomers.

Table 3 Propagation and Termination Rate Constants of Radical and Ionic Polymerising Monomers

Monomer	Initiation by	Temperature	k_p (l mol^{-1}s^{-1})	k_t (l mol^{-1}s^{-1})
Methyl methacrylate [20]	radicals	30	286	2.4x10^7
Styrene [20]	radicals	25	39.5	5.96x10^6
Polyurethane acrylate +hexanedioldiacrylate [21]	radicals	RT		5x10^5
Polyurethane acrylate +hydroxyethyl-carbamate acrylate [21]	radicals	RT	1x10^4	
Styrene [20]	cations	15	3.4x10^6	2.3x10^{11}
Isobutene [20]	cations	0	1.5x10^8	9.3x10^{11}
Vinylethyl ether [20]	cations	30	9.4x10^4	3.5x10^{11}
α-Methyl styrene [20]	cations	30	3x10^6	8.4x10^9

(iii) Effect of oxygen in electron beam curing

It is well known that oxygen inhibits electron beam curing of acrylates and other monomers polymerising by a radical mechanism. In practical curing applications, nitrogen blanketing is used to prevent the direct contact of oxygen with the coating to be cured. Oxygen undergoes reactions with various radicals formed after electron irradiation of vinyl monomers. The oxygen concentration within the coating considerably changes during curing. Oxygen consumption by radicals and oxygen diffusion has to be taken into account. As shown in Table 2 of this Chapter, oxygen can diffuse over microns during 100 ms in low viscosity liquids. Hence, the oxygen effect on electron beam curing of coatings has to be considered as an interplay of oxygen depletion by chemical reactions and oxygen diffusion into or out of the coating.

In order to illustrate the effect of oxygen on electron beam initiated radical curing, the mechanism of radical formation in acrylates is used as an example. As shown in Chapter 1, Figure 14 and schematically outlined in Figure 28, fast electrons first generate the primary ion pair consisting of an acrylate radical cation (Ac^+) and a solvated electron (e_s^-).

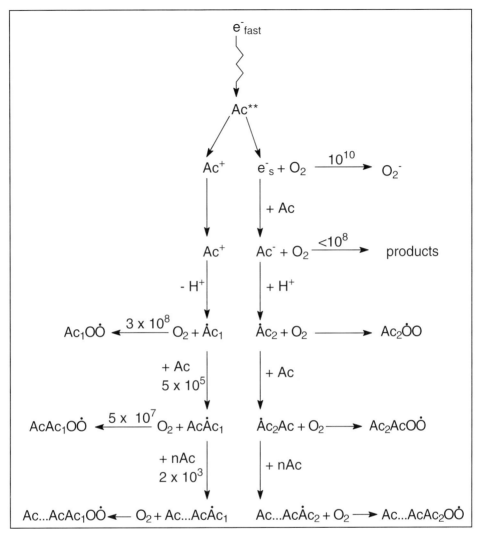

Figure 29 The Effect of Oxygen on Various Reaction Steps in Electron Beam Curing of Acrylates

As pulse radiolysis experiments on TPGDA show [22], the cation is insensitive to oxygen but electrons usually react with oxygen at diffusion controlled rate constants. However, acrylates are also good electron scavengers. A reaction rate constant of 3×10^{10} l mol^{-1}s^{-1} was determined for electron attachment to TPGDA. Even at diffusion-controlled reaction of the electron with oxygen (oxygen saturation concentration in TPGDA $\approx 2 \times 10^{-3}$ mol^{-1}), due to the high TPGDA concentration of 3.5 mol^{-1}, acrylate anion formation is by far faster. Thus, oxygen does practically not interfere in anion formation.

Acrylate anion radicals also react with oxygen. In oxygen saturated solutions this reaction can even compete with anion protonation, decreasing the concentration of initiating radicals generated via the anionic pathway.

However, oxygen has a more pronounced effect on chain initiation. As shown in Figure 29, TPGDA peroxy radicals are formed with a rate constant of 3×10^8 l mol^{-1} s^{-1}. At oxygen saturation, a reaction rate (2×10^{-3} mol^{-1} $\times$ 3×10^8 mol^{-1}s^{-1}) of 6×10^5 s^{-1} is obtained, which compares with that of 17.5×10^5 s^{-1} estimated for chain initiation.

Chain propagation is also strongly affected by oxygen. A propagation rate of 7×10^3 s^{-1} has to be compared with an oxygen termination rate of 1×10^5 s^{-1}. As a result, polymerisation should not take place in the presence of oxygen.

This is not completely in accordance with the experience in EB curing applications. If nitrogen blanketing is not used, a tacky surface is obtained in all cases, whereas for thicker coatings often some bottom cure is observed (see Figure 32b of this Chapter). Obviously, in this thickness range oxygen depletion by chemical reactions cannot be compensated by oxygen diffusion from the surrounding air. For electron beam curing, calculation of the reaction kinetics under diffusional motion is complicated by the changes in the diffusion coefficient during network formation, and is thus not tried here. However, an example may illustrate the effect of combined oxygen depletion and regeneration by diffusion:

First we use Eqn. (18) $v_i = G_r D° \rho$ to estimate the initiation rate constant. Assuming a radical G-value G_r of 0.5×10^{-7} mol J^{-1} and a dose of 5×10^4 G$_y$ (defined as J kg^{-1}) delivered during an irradiation time of 100 ms and a density $\rho = 1$ kg l^{-1}, the following initiation rate constant is obtained:

$$v_i = 0.5 \times 10^{-7} \times 5 \times 10^5 = 2.5 \times 10^{-2} \text{ mol l}^{-1}\text{s}^{-1},$$

i.e. during the irradiation time of 100 ms the oxygen saturation concentration of 2×10^{-3} mol l^{-1} is completely depleted by chemical reactions.

Now oxygen diffusion is taken into account as shown in Table 2, the mean oxygen diffusion distances calculated for low viscosity liquids amount to between 14.1 and 4.5 μm within 100 ms, i.e. oxygen diffusion nearly compensates for oxygen depletion by chemical reactions. In this thickness range the coating remains uncured.

If nitrogen blanketing is used, the tolerable residual oxygen concentration is of great practical importance. It determines the necessary technical standard of the nitrogen inerting system and the nitrogen flow rate. The residual oxygen concentration tolerable for electron beam curing of acrylates can be estimated using the rate constants of the most important competing processes: chain initiation and propagation vs. peroxy radical formation.

Using as rate constants:

$k_i = 5 \times 10^5$ l mol^{-1}s^{-1} and $k_{iox} = 3 \times 10^8$ lmol^{-1}s^{-1} for the initiation,

$k_p = 2 \times 10^4$ l mol^{-1}s^{-1} and $k_{pox} = 5 \times 10^7$ lmol^{-1}s^{-1} for the propagation, and

the monomer concentration [M] = 3.5 mol^{-1},

the oxygen concentration in the coating [O$_2$] must be:

$$[O_2] < k_p/k_{pox}[M] = 1.32 \times 10^{-3} \text{ moll}^{-1} = 0.042 \text{ gl}^{-1} = 45 \text{ ppm}$$
$$[O_2] < k_i/k_{iox}[M] = 5.5 \times 10^{-3} \text{ moll}^{-1} = 0.176 \text{ gl}^{-1} = 187 \text{ ppm}.$$

This example shows at least the order of magnitude which the oxygen concentration in the coating must reach before effective curing can occur.

On the other hand, this implies that a residual oxygen concentration of about 200 ppm in the inert gas does not affect curing. With the exception of silicon acrylates, which for curing need oxygen concentrations of less than 50 ppm, 200 ppm oxygen is tolerable for electron beam curing of acrylates.

(iv) Pigmentation in electron beam curing

As mentioned in Chapter I, fast electrons are absorbed according to the electron densities of the absorbers in the irradiated matrix. In the characteristic dose vs. penetration depth profiles of electrons the dose is given as a function of the aerial density. Hence, the penetration depth of electrons of a certain energy is only affected by the density of the matrix to be irradiated. Chromophoric effects as observed after photon absorption are completely absent.

II. DEGREE OF CURE: PHYSICAL AND CHEMICAL CHARACTERISATION

1. Methods to measure the degree of cure

When operating with the definition of degree of cure that

the cured product meets the needs of its function properties to be commercial

it is important to have meaningful tests which will ensure the quality of the cured product and reproducibility of the curing process.

There are many different methods for evaluating the degree of cure. Some of them are strictly laboratory others are field tests. However, the converter of UV(EB) products is most interested in practical field tests that allow an evaluation of product quality during or immediately after the run. On the other hand, formulating of UV(EB) coatings inks and paints, and curing technology development call for more precise laboratory test methods. For this reason both, simple field and more sophisticated laboratory test will be described here.

(i) Field cure tests

All tests summarised in Table 4 are practical tests that can be performed during or immediately after curing [23].

In some cases, not even auxiliary devices are needed for such simple tests. Although the degree of cure cannot be determined precisely, the practical test allows the end user of UV(EB) curing technologies to be certain that the products will meet his customers' needs.

When tests can be correlated with desired end properties, this aim is achieved.

Table 4 Field Cure Tests

Test	Characteristics	Reproducibility affected by	Remarks
Solvent resistance	A solvent containing wet cloth is rubbed forth and back on the sample. Double rubs are counted until solvent breaks through the coated layer. Methyl ethyl ketone (MEK) is often used as a solvent.	nature of coating, solvent used, pressure and speed of rubbing, coating thickness	Cure analyser commercially available that measures the evaporation rate of a solvent placed on the sample.
Stain resistance	A coloured solution, e.g. 1% potassium permanganate in water, is allowed to stand on the cured coating for a defined time. When the solution is wiped off the resultant stain is evaluated.	colour of the coating, temperature, formulation of the coating, surface roughness of the coating.	This test measures surface rather than through cure.
Hardness test pencil	The hardness is determined by a set of numbered pencils. The number of the first pencil which scratches the surface is taken as hardness of the coating.	temperature, applied pressure, film thickness, substrate.	ASTM D 3363 Calibrated set of pencils: 6B-5B-4B-3B-2B-1B-HB-F-H-2H-3H-4H-5H-6H, Hard
pendulum (Koenig, Persoz)	Pendulum hardness is determined by the attenuation of pendulum oscillation by a coating on a glass plate. The damping time required to slow the oscillation of the pendulum down from 6° to 3° (Koenig) or from 12° to 6° (Persoz) is measured.	temperature	DIN 53 157 SNV 37 112
Abrasion resistance Thumb	The sample is placed on a solid surface and pressed with the thumb. With a high pressure the operator quickly twists his thumb. The	pressure, operator.	

twist	sample is then examined for deformation.		
Taber	Abrasion is produced by contact of the test sample, turning on a vertical axis, against the sliding rotation of two abrading wheels. Weight loss of the sample is determined after a certain number of cycles. Depth of wear can also be determined.	temperature, slip of the sample,	Taber abraser commercially available. DIN 53 754, DIN 52 347 ISO /DIS 7784.2
Tack test	Sticking of a cotton ball on the coated surface is evaluated.	many aspects.	Decision between cured and uncured only

(ii) Laboratory cure test methods

In contrast to field cure test methods, in most cases laboratory cure test methods are sophisticated analytical techniques used to study kinetics and degree of monomer conversion or polymer formation. Depending on their mode of operation, analytical cure test methods can be divided into two classes: methods based on discrete measurements of monomer conversion after discrete reaction steps and methods allowing a real-time monitoring of curing. In Figure 30 stationary and time-resolved methods for assessing the degree of cure and cure kinetics are summarised and the physical or chemical properties correlated to monomer conversion are indicated.

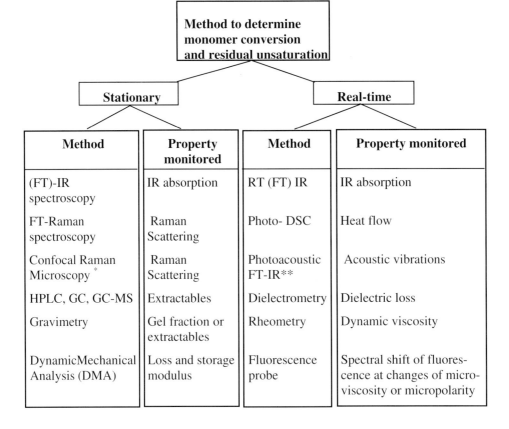

* Depth profiling and lateral mapping of the degree of cure possible
** Depth profiling possible

Figure 30 Laboratory Cure Test Methods

A complete technical description of all the analytical methods mentioned in Figure 30 is not the purpose of this chapter (for more information see, e.g. references (1), (3) and (4))but the following brief description of each method concentrates on the basic measuring principle and the specific information gained by the method.

(a) *Infrared and Raman spectroscopy*

Infrared (IR) spectroscopy is widely used to study UV (EB) curing of a large variety of radiation-curable resins, simply by monitoring molecular vibrations correlated to the monomer reactive moiety. In a similar way, Raman spectroscopy is used. Infrared absorption is observed when the dipole moment of the molecule changes during the vibration, whereas Raman scattering is observed when polarisibility changes.

In this respect, IR and Raman spectroscopy are complementary methods to determine monomer conversion by vibrational spectroscopy.
IR and Raman spectroscopy provide quantitative results on both the degree of conversion and the polymerisation rate. After curing has finished, the amount of residual unsaturation can precisely be determined from the remaining absorbance.

Vibrational bands which can be used to measure acrylate, epoxide or vinyl ether conversion, are summarised in Table 5. In many cases it is very useful to normalise the vibrational absorption to an internal standard. This is a vibrational band that does not change its intensity during curing. Thus, spectra measured at different irradiation doses can be precisely compared.

Table 5 Vibrational Bands Used to Monitor Monomer Conversion in UV (EB) Curing*

* D. Lin-Vien, N.B.Colthup,W.G. Fateley,J.G. Grasselli: The Handbook of Infrared and Raman, Characteristic Frequencies of Organic Molecules, Academic Press, San Diego, 1991, N.B.Colthup, L.H.Daly, S.E.Wiberley: Introduction to Infrared and Raman Spectroscopy, Academic Press, San Diego, 1990 (3rd ed.)

Vibrational band of the monomer (cm^{-1})	Assignment of vibration	Detected by	Reference band used as internal standard (cm^{-1})
810 acrylate	CH=CH$_2$ twisting	IR	CH$_2$ stretching 2862
1190 acrylate	C–O stretching	IR	
1410 acrylate	CH$_2$ sci.deformation	IR/Raman	C=O stretching 1720-1730
1639 acrylate	CH=CH$_2$ streching	IR/Raman	
790 epoxide	Ring deformation	IR/Raman	
1110 epoxide	C–O–C as. stretching	IR	
1616-1622 vinyl ether	CH=CH$_2$ stretching	IR/Raman	

Figure 31 shows, as examples, IR spectra of our model system TPGDA/1 wt. IC369 and of the cationic system UVR-6105/Tone 0301/UVI-6990 before and after UV exposure. The decrease of the peak areas at 810 cm^{-1} or 1639 cm^{-1} for the acrylate and at 790 cm^{-1} for the epoxide system exactly reflects the degree of monomer conversion.

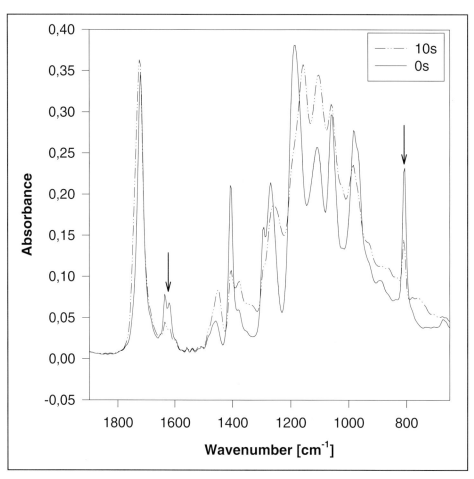

Figure 31a Infrared Absorption Spectrum of TPGDA/ 1 wt.-% IC 369 Measured Before (0 s) and After 10 s of Irradiation with 313 nm Light, Irradiance 48 mWcm^{-2}

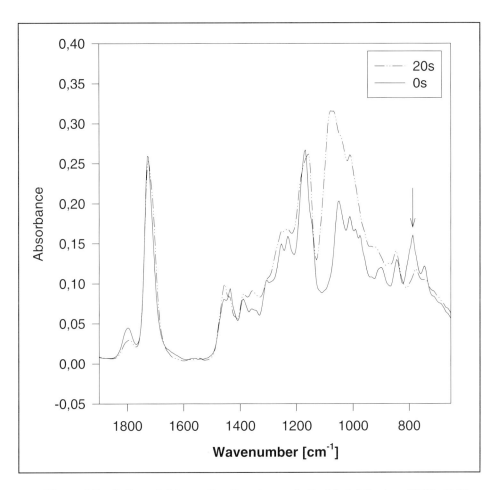

**Figure 31b Infrared Absorption Spectrum of the Model System UVR- 6105
(75 wt.-%), Tone 0301 (25 wt.-%) and 6 wt.-% UVI-6990
Before (0 s) and After (20 s) Irradiation with 313 nm Light, Irradiance 100 mWcm^{-2}**

The degree of conversion x obtained after a given time t can be directly calculated from:

$$x = 1 - A_t/A_0, \tag{20}$$

where A_0 and A_t are the peak areas (corresponding to absorbances) measured for the characteristic vibrations before and after exposure, respectively. From conversion vs. time profiles, as obtained by plotting the degree of conversion x as a function of time, the polymerisation rate v_p can be calculated. Additionally, when conversion is finished at long exposure times, from the remaining peak area the residual unsaturation content can be estimated.

Measuring conversion vs. time profiles by discrete steps using stationary infrared or Raman spectroscopy is time consuming and often not very accurate. An important technical advantage was obtained by C. Decker and K. Moussa [5,24,25], who combined in-situ UV irradiation of the sample with real-time infrared monitoring of reactive monomer bands.

(b) Real-time infrared spectroscopy

The first real-time IR spectrophotometer developed by C. Decker was used in the absorbance mode at a fixed wavelength. The sample was placed on potassium bromide crystal plates and exposed to UV light. Simultaneously to irradiation, the decrease of a reactive monomer band was monitored in real-time mode.

After more than 10 years, highly sensitive FT IR spectrometers are now available which are equipped with attenuated total reflectance (ATR) or reflection-absorption (RA) units. Typically 4-100 spectra can be recorded during 1 s.

The experimental set-up of an ATR-Fourier Transform Infrared spectrometer is given in Figure 1 of this chapter [26]. In a similar way, an FTIR spectrometer equipped with a large angle reflectance infrared (LARI) cell can be used [27].
This makes real-time IR spectroscopy a very versatile technique for UV curing kinetics studies:

- A time resolution of 10 ms is reached, i.e. 100 full spectra are recorded within 1s.
- Samples can now be placed horizontally onto the ATR crystal
  or the LARI sample holder.
- The temperature of the sample can be varied up to 200°C.
- Purging with inert gas is possible.
- Pigmented samples and powders can be analysed.

All the examples shown in this chapter in order to illustrate the UV curing kinetics have been obtained by using the real-time FTIR-ATR technique.

(c) Confocal Raman spectroscopy

Confocal microscopy [28] uses laser excitation of the sample and observation of a confocal volume of typically 1 µm^3 within the sample. The volume to be observed is selected by focusing the confocal aperture microscope objective to the sample volume position or by moving the sample through the focal point. Thus, spatial resolution in the range of µm is obtained.

If the laser applied as excitation source in confocal microscopy can also be used to induce Raman scattering, Raman spectroscopy with micrometer spatial resolution becomes possible [29].

Confocal Raman microscopy combines both, molecular information from vibrational spectroscopy and spatial resolution from confocal microscopy.
Figure 32 shows the scheme of a confocal Raman spectrometer [30].

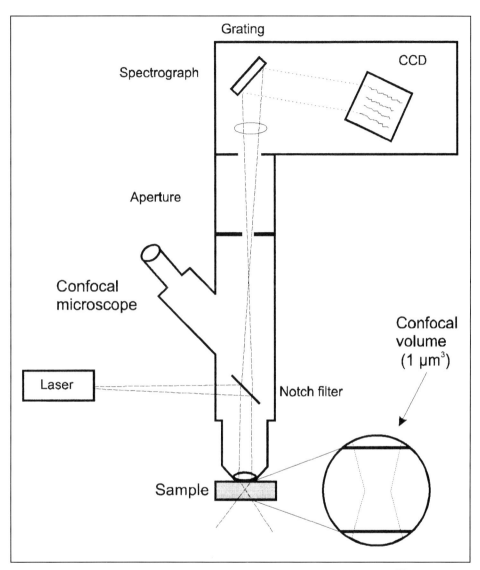

Figure 32a Scheme of a Confocal Raman Spectrometer [30]

Unlike the very small sample volume, moderate measuring times are obtained. This is due to improvements in recording of scattered Raman photons: a special notch filter suppresses laser scattering while allowing transmission of Raman photons, a bright single grating spectrograph and a sensitive low-noise (cooled) CCD array are used as detection equipment.

In UV(EB) curing the determination of the degree of cure as a function of the penetration depth of radiation is of great importance. Confocal Raman spectroscopy offers a unique opportunity to avoid, for example, time consuming microtome cutting and to determine cure depth profiles, simply by taking Raman spectra as a function of depth. Figure 32b shows the depth profiles measured in a polyether acrylate sample which was cured under air. As a result, the sample surface remained tacky. This is clearly reflected by the absence of any conversion within a surface layer of 5 µm thickness.

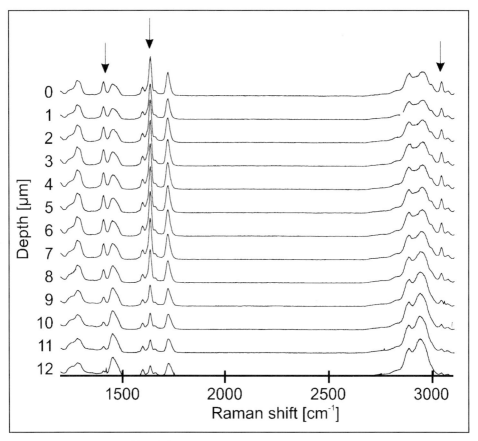

Figure 32b Depth Profile of Curing of Polyetheracrylate. Reactive Acrylate Groups are Marked by Arrows

Confocal Raman spectroscopy opens up the new and unique field of non-destructive optical detection of spatial concentration gradients which are of importance in UV(EB) curing. Typical problems which can be investigated are:
- degree of cure as a function of depth,

- cure inhibition as a function of depth,
- lateral distribution of UV absorbers as a function of depth,
- effect of pigmentation in UV(EB) curing.

(d) Photo- Calorimetry (DSC)

Since its introduction by Moore [31], photo-isothermal differential scanning calorimetry (DSC) has become the most common method used for real-time monitoring of UV curing.

During UV exposure a differential scanning calorimeter measures both the heat flow and the total amount of heat evolved. The latter is related to the final degree of cure while heat flow is a measure of the polymerisation rate. This method allows time resolutions of seconds and has been applied to extensively study the effect of chemical and physical factors on the polymerisation rate and cure kinetics such as monomer or oligomer functionality, photoinitiator action, oxygen content, UV source irradiance and spectrum, temperature, postcuring, induction period, sample thickness and others.

In isothermal DSC the heat flow, commonly expressed in Wg^{-1} or $J\ g^{-1}\ s^{-1}$, is recorded as a function of time. Figure 33 shows typical photo-DSC profiles obtained by exposing an acrylate and an epoxide formulation to light of water-filtered radiation from a medium pressure mercury lamp.

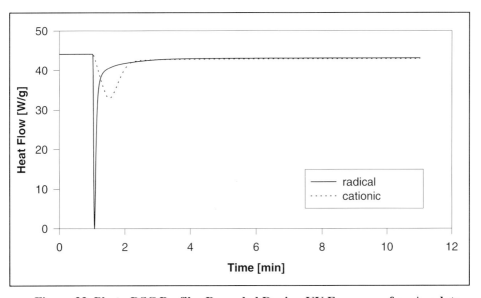

Figure 33 Photo-DSC Profiles Recorded During UV Exposure of an Acrylate (Radical) and an Epoxide (Cationic) System

The rate of polymerisation v_p can be calculated from heat flow (dH/dt) and the molar standard heat of polymerisation of the monomers ΔH_0 used:

$$v_p = (dH/dt)_{\text{J mol}^{-1}\text{s}^{-1}} \times \frac{[M_0]_{\text{mol l}^{-1}}}{\Delta H_{0\ \text{J mol}^{-1}}}, \qquad (21)$$

where $[M_0]$ is the initial monomer concentration.

Molar heats of polymerisation measured for reactive groups of several UV curable monomers are listed in Table 6. It is commonly assumed that the molar heat is not greatly affected by substituents which are not directly bound to the reactive group.

Table 6 Molar Heat of Polymerisation of Reactive Groups

Reactive group	Heat of polymerisation (kJmol^{-1})
Acrylate	78 - 86
Methacrylate	57
Vinyl ether	60
Oxirane ring	88
Vinyl acetate	80

Unlike its wide use, Photo-DSC results are not always comparable to those obtained under practical UV curing conditions. An important limitation is the low thermal conductivity of the sample and the slow time response of DSC detection. Even response times of seconds imply that low irradition intensities have to be applied, usually up to two orders in magnitude lower than in practical UV curing systems. Additionally, the minimum sample weight of 1 mg leads to a sample thickness of about 50 μm when standard sample pans are used. It is very difficult to analyse thinner samples by Photo-DSC, but in practical curing applications thinner coatings dominate.

All limitations mentioned for Photo-DSC are not valid for real-time FTIR. In that case, response times of 10 ms are possible and even pigmented samples of 1μm thickness can be analysed with high sensitivity. All the physical and chemical effects on the polymerisation rate and the degree of cure can be studied as well. Hence, modern real-time FT IR measurements of curing kinetics are in most cases superior to those done by Photo-DSC.

(e) Fluorescence probe technique

Fluorescent compounds which display a shift of their emission spectrum with changes of microviscosity or micropolarity are suitable for application as polymerisation fluorescence probes. When a small amount of such a probe is added to monomers, characteristic changes of the fluorescence emission can be observed during progressing conversion. Neckers and his group ([32,33,34]) have identified a number of fluorescent compounds which show a blue shift upon polymerisation.

The most commonly used fluorescence probe is 5-dimethylamino naphthalene-1-sulphonyl-n-butylamide (1,5-DASB). Figure 34 shows emission spectra and structure of 1,5-DASB.

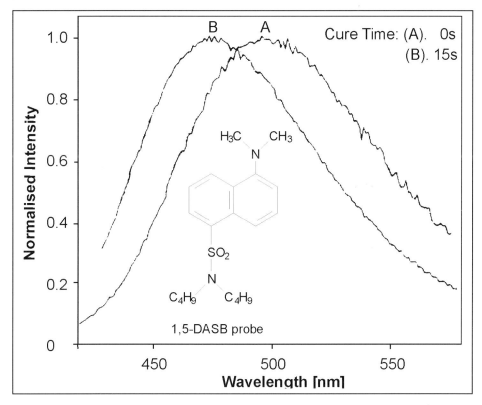

Figure 34 Fluorescence Spectrum of 1,5-DASB at Different Degrees of Cure

Figure 34 shows that a small blue shift occurs upon polymerisation. To quantify this effect, the fluorescence intensity ratio of two wavelengths is calculated. With progressive curing the fluorescence intensity on the left side of the spectrum increases, while the intensity on the other side decreases. Thus, the fluorescence intensity ratio increases with increasing degree of cure. Calibration of the fluorescence intensity ratio vs. degree of cure can be done by using a set of cured samples with a known degree of cure, determined, for example, by stationary FTIR spectroscopy. Recently, a rapid scan fluorimeter has been modified to allow sample excitation and fluorescence detection by a flexible fibre optics sample head [35,36].

This "cure monitor" acquires a full fluorescence spectrum within 0.1 to 0.25 s. As excitation range of 300 to 400 nm was selected, while fluorescence probe emission is observed between 400 and 700 nm. Unlike off-line cure monitoring, the fluorescence probe technique also offers some potential for on-line monitoring. Additionally, the excitation light can be used to induce fluorescence excitation as well as curing. In this way, in situ cure monitoring becomes possible.

Another interesting fluorescent probe system has been investigated by Warman et al. [37]. They use a fluoroprobe covalently bound to maleimide, which fluoresces only when it is copolymerised into a polymer chain. Thus, the fluorscence intensity is directly related to the number of fluoroprobes incorporated into the polymeric network.

(f)   Photoacoustic FT IR spectroscopy

The degree of cure and cure kinetics of certain UV cured samples cannot be analysed by routine IR transmission or reflection techniques. Examples of such samples are coatings on fibres, composites and coatings containing micro- or nanoparticles in a non-homogeneous distribution and/or unusual shape, foams, silica gel etc. Photoacoustic spectroscopy is a technique which permits non-destructive analyses of such samples with no or minimum sample preparation [38]. To some extent, even depth profiling is possible [39].

Photoacoustic FTIR spectroscopy is based on the effect that the energy absorbed by the characteristic vibrational frequencies of the sample is converted into heat. As Figure 35 shows, this heat will cause the gas on the sample`s surface to warm up and expand.

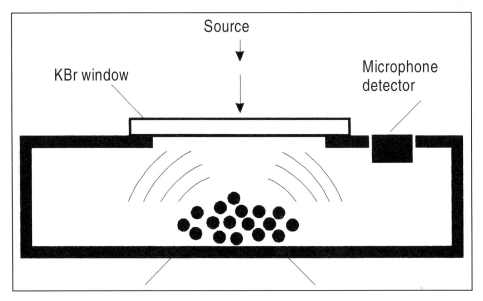

Figure 35 Photoacoustic Detector Cell

In a sealed volume this creates an increase in pressure which can be detected as an acoustic signal by a microphone. Because the IR beam of a FTIR spectrometer is modulated by a Michelson interferometer, the microphone signal is obtained as interferogram. As in the case of direct infrared detection, the interferogram can be mathematically transformed to give a spectrum. The sample spectrum is usually related to the carbon black spectrum, which absorbs all wavelengths. Because the photoacoustic effect is caused by heating, the sample surface and the thermal penetration depth L is determined by:

$$L = (D/\pi v f)^{1/2}, \qquad (22)$$

where D is the thermal diffusivity of the sample material, v is the scanning velocity of the spectrometer and f is the infrared wavenumber. Depth profiling is possible by changing the scan speed. Figure 36 shows the optical scheme of a photoacoustic accessory to a FT IR spectrometer.

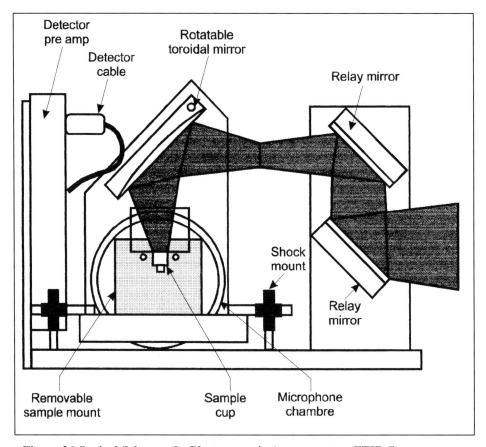

Figure 36 Optical Scheme of a Photoacoustic Accessory to a FTIR Spectrometer

(g)   Gel Content and Extractables

During UV curing of multifunctional monomers network formation takes place by polymerisation and cross-linking. When monomer conversion exceeds the gelation point, two macromolecular systems coexist:
- a sol consisting of non-crosslinked polymer chains, which is soluble and
- a gel composed of cross-linked polymer chains, which is insoluble even in good solvents.

One of the simplest methods to determine the degree of cure consists in weighing the amount of insoluble material formed after UV exposure.

The gel content is commonly measured by gravimetry after Soxhlet extraction and drying of the sample. Methyl ethyl ketone is often applied as a solvent. Typically, the dispersed sample is refluxed for two hours. After filtration and evaporation of the solvent the gel content G is determined as:

$$G = \frac{w_g}{w} \times 100 \; (\%) \qquad (23)$$

where w is the weight of the cured sample before extraction and w_g the weight of the insoluble residue.

On the other hand, the extract contains soluble components from the cured sample such as residual monomers, oligomers, additives, photoinitator and photoinitator products. Nature and quantitiy of the products solved can be determined using common analytical methods such as HPLC, GC or GC-MS. The total amount of extractables obtained from a cured sample is an important quantity to qualify the extent of cure. This is especially the case when packaging material for foodstuffs or consumer goods is evaluated. For this case, model procedures were worked out which are in accordance with European and US regulations [40, 41].

Test simulants for foodstuffs are defined: water 3 wt.-% of acetic acid, water 15 per cent by volume of ethyl alcohol, olive oil or some equivalent fat, which, in an especially designed cell, have to be kept in direct or indirect contact with the cured material for 1 to 10 days. Much faster extraction (minutes) can be achieved when an ultrasonic bath with acetonitrile as solvent is used. For aqueous and acetonitrile solutions HPLC, and for more volatile components, GC and GC-MS are methods of choice to determine, for example, photoinitiators, their decomposition products and monomers.

Figure 37 shows as an example the gel content of an acrylate sample determined as a function of the EB dose and the corresponding amount of extractables as measured by HPLC.

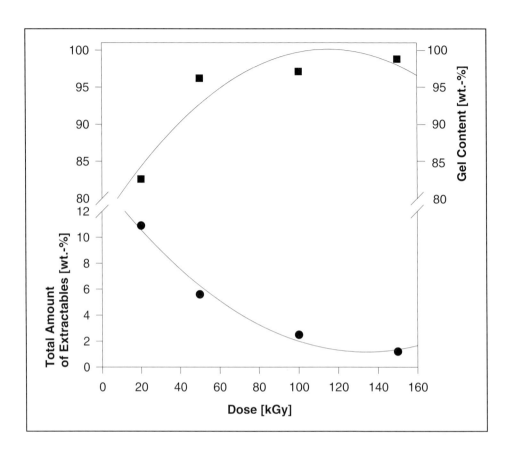

Figure 37 Gel Content and Total Amount of Extractables as Function of Dose. System: Electron-Irradiated Acrylate Formulation

(h) Dynamic Mechanical Analysis (DMA) and Rheometry

The DMA technique measures viscoelastic properties which are related to molecular motion in polymers. A small sinusoidal varying stress (force per area) is exerted on the polymer material under test and the resulting strain (e.g., elongation per original length) response is recorded [42, 43].

For an ideal elastic material the strain response is in phase with the stress. With increasing viscosity the strain lags in phase behind the stress.

The viscoelastic properties of polymers are dependent on temperature. Thus, DMA profiles commonly display an elastic-like component, expressed by the storage modulus, and a viscous-like component, governed by the loss modulus, as a function of temperature.

It is obvious, that DMA offers a useful qualitative method to study the transformation of a fluid film into a solid material through UV exposure or electron irradiation. The degree of cure can be determined by measuring the changes of storage and loss modulus as a function of the exposure time (or UV/EB dose). Figure 38 shows DMA profiles measured for electron-irradiated TPGDA at three different doses.

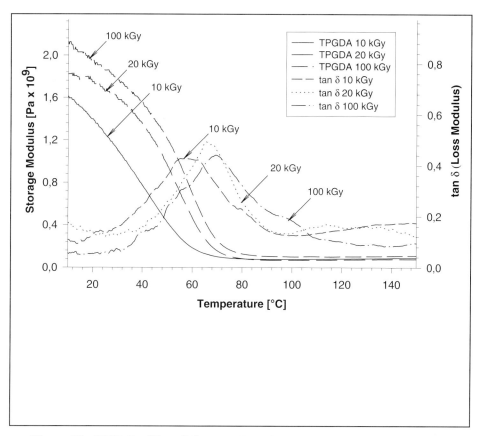

Figure 38 DMA Profiles of Electron-Irradiated TPGDA as Function of Dose

A main limitation of this method is that the DMA data cannot be directly related to monomer conversion. Thus, the method allows qualitative results only.

In this respect, DMA is similar to methods which measure changes in dynamic viscosity or rheology upon UV exposure. Oscillating plate rheometers or vibrating needle devices rather measure gelation than the full degree of cure.

(i) Dilatometry

Dilatometry is a standard technique to follow the volume change, i.e. shrinkage, during polymerisation. During polymerisation, all vinyl monomers show a relatively large amount of shrinkage. For acrylates and methacrylates shrinkage can reach even 20%.

Although shrinkage is not a direct measure of monomer conversion, there is a dependence of shrinkage on conversion.

The dilatometric technique can also be applied to studying UV curing of thin samples in real time [44].

In this case, the sample is UV cured in a closed system filled with water having a capillary glass tube as exit. When shrinkage occurs, water in a glass capillary tube changes its filling position. The capillary tube is placed between capacitor plates. The capacitance of the arrangement is dependent on the position of the air-water interface. Thus, shrinkage changes are transformed to changes in capacity, which are recorded as a function of time.

III. REFERENCES

1. S.P. Pappas (ed.) : C. Decker in:Radiation Curing Science and Technology, Plenum Press, New York and London , p.135 (1992)
2. P.K.T. Oldring (ed.): Chemistry &Technology of UV and EB Formulation for Coatings, Inks & Paints, Vol. 2 and 3, SITA Technology, London, (1991)
3. J.P. Fouassier and J.F. Rabek (ed.): J.F.Rabek in: Radiation Curing in Polymer Science and Technology, Vol. I and II, Elsevier Science Publishers, London and New York, (1993)
4. Ch.E. Hoyle and J.F. Kinstle (eds.) : C. Decker and K. Moussa in : Radiation Curing of Polymeric Materials, ACS Symposium Series, Washington, p. 439 (1990)
5. C. Decker, K. Moussa, Makromol. Chem. 189, p. 2381 (1988)
6. A.J. Bean, Proceedings RadTech North America `94, p.725 (1994)
7. C. Decker, B. Elzaouk, D. Decker, Proceedings RadTech Europe`95, Academic Day, p.115 (1995)
8. J. Wendrinski, R. Liska, Proceedings RadTech Europe `97, p. 569 (1997)
9. K.K. Dietliker in: P.K.T. Oldring (ed.): Chemistry&Technology of UV and EB Formulation for Coatings, Inks&Paints, Vol. 3, SITA Technology, London, (1991)
10. G.R. Tryson, A.R.J. Shultz, J. Polym. Sci. Polym. Phys. Edn. 17, p. 2059 (1979)
11. U. Decker, private communication
12. R. Mehnert, Farbe und Lack, 100 (5), p. 325 (1994)
13. T. Scherzer, U. Decker R. Mehnert, Proceedings RadTech`98 North America, p.746 (1998)
14. , R.W. Stowe, Proceedings RadTech North America`90, p.173 (1990)
15. D. Skinner, Proceedings RadTech Europe`97, p.125 (1997)
16. J.V. Crivello, D.A.Conlon, D.R. Olson, K.K. Webb, Proceedings Radcure Europe`87, p. 1-27 (1987)
17. J.V. Koleske, Proceedings RadTech North America`88, p.353 (1988)
18. S.C. Lapin, J.R. Snyder, Proceedings RadTech North America `90, p. 410 (1990)
19. C. Priou, J.M. Frances, Proceedings RadTech Europe`97, p. 314 (1997)
20. V.S. Ivanov: Radiation Chemistry of Polymers, VSP Utrecht, 1992
21. C. Decker, B. Elzaouk,D.Decker, Proceedings Radtech Europe`95, Academic Day, p. 115 (1995)
22. W. Knolle, R. Mehnert, Radiat.Phys.Chem. 46, p. 963 (1995) and Nucl.I nstr.and Meth. B 105, p. 154 (1995)
23. C.J. Callendorf: Radiation Curing Test Methods, RadTech International, (1988)
24. C. Decker, K. Moussa, J. Coatings Technol. 62 , p. 55 (1990)
25. C. Decker, K. Moussa, J. Polym. Sci. Chem. Ed. 28, p. 3429 (1990)
26. T. Scherzer, U. Decker, Vibr. Spectroscopy, to be published
27. A.A. Dias, H. Hartwig, J. Jansen, Proceedings RadTech `98 North America , p.356 (1998)
28. T. Wilson: Confocal Microscopy, Academic Press, London, (1990)
29. B. Schrader: Infrared and Raman Spectroscopy, VCH, Weinheim, (1995)

30. W. Schrof, L. Häußling, R. Schwalm, W. Reich, K. Menzel, R. Königer, E. Beck, Proceedings RadTech`97 Europe, p. 535 (1997)
31. J.E. Moore in S.P. Pappas (ed.), UV Curing: Science and Technology, Vol.1, p.134, Technology Marketing Corp., Norwalk, (1978)
32. AJ. Paczkowski, D.C. Neckers, Macromolecules, 24, p. 3013 (1991)
33. A.J. Paczkowski, D.C. Neckers, Macromolecules, 25, p. 548 (1992)
34. J.C. Song, A. Torres-Filho, D.C. Neckers Proceedings RadTech `94 North America, p. 338 (1994)
35. R. Popielarz, D.C. Neckers, Proceedings RadTech`96 North America, p.271 (1996)
36. K.G. Specht, R. Popielarz, S. Hu, D.C. Neckers, Proceedings RadTech`98 North America, p. 348 (1998)
37. J.M. Warman, R.B. Abellon, H. Verhey, J.W. Verhoeven, J.W. Hofstraat, J. Phys. Chem. 101, p. 4913 (1997)
38. P.R. Griffiths, J.A. de Haseth: Fourier Transform Infrared Spectroscopy, J. Wiley Interscience, NewYork, (1986)
39. R.W. Jones, J.F. McClelland, Appl.Spectroscopy,50, p. 1258 (1996)
40. R. Derra, Proceedings RadTech Europe`97, p. 583 (1997)
41. I.D. Newton in Polymer Characterisation, B.J. Hunt and M.I. James (eds.), Blackie Academic, London, p. 8–36 (1993)
42. M.E. Brown: Introduction to Thermal Analysis, Chapman and Hall, London, New York, (1988)
43. R.E. Wetton, in Polymer Characterisation, B.J. Hunt and M.I. James (eds.), Blackie Academic, London, p.178-221 (1993)
44. R.B. Cundall, Y.M. Dandiker, A.K. Davies, M.S. Salim in Radiation Curing of Polymers, D.R. Randall (ed.), Special Publication N0 64, Royal Soc. Chem., London, p. 172-183 (1987)

CHAPTER IX

UV&EB EQUIPMENT HEALTH AND SAFETY

1. UV Equipment Health and Safety

There is no doubt that the UV curing technology has contributed significantly to the world-wide reduction in VOC (volatile organic compound) emission. Although this technology is regarded as inherently environmentally-friendly, there are health and safety issues related to the formulation, coating and curing process and the UV equipment. Some of these issues relate to the chemical structure and to physico-chemical properties of monomers and oligomers used in formulations of UV curable inks, paints, varnishes, adhesives, functional coatings etc. Health and safety instructions for these specific chemical products are provided by the manufacturer.

In this context health and safety aspects will be discussed which relate to the specific field of operation of UV curing equipment.

UV curing systems have to be designed to suit their application, but also to provide optimum protection of the operator against:
- scattered UV radiation and heat,
- ozone generation and
- high voltages.

(a) Photobiological effects of UV radiation

Exposure to UV radiation may cause several photobiological effects such as reddening of the skin (erythema), skin burning, premature skin ageing, skin pigmentation but also damage to eyes such as photo-conjunctivitis and -ceravitis. Irritations of the eye which last for hours or days and then disappear with no permanent damage may also be caused. Snow-blindness or welder´s flash are such transient effects.

Absorption of UV radiation can lead to photoinduced chemical reactions which finally cause the specific photobiological effect. The photobiologically efficient irradiance E_{biol} is defined by the following integral:

$$E_{biol} = \int_{\lambda 1}^{\lambda 2} E_\lambda(\lambda)\, s_{biol}(\lambda)\, d\lambda, \tag{1}$$

where E_λ is the spectral irradiance and $s_{biol}(\lambda)$ the spectral sensitivity related to a specific photobiological effect. The magnitude of the specific photobiological effect H_{biol} is determined by the product irradiance E_{biol} times exposure time t:

$$H_{biol} = \int_0^t E_{biol}(\lambda)\, dt. \tag{2}$$

According to its definition H_{biol} is a UV dose measured in $mJ\ cm^{-2}$. The dose causing a detectable photobiological effect is defined as the threshold dose $H_{th,\ biol}$.

Table 1 summarises threshold doses $H_{th,\ biol}$ of different photobiological effects and the wavelengths of maximum receptor sensitivity.

Table 1 Photobiological Effects of UV Irradiation (According to DIN 5031-10, Draft Document 17-96)

Photobiological Effect	Threshold Dose $H_{th, biol}$ [mJcm^{-2}]	Wavelength of Maximum Receptor Sensitivity [nm]
UV Erythema	20 - 45	298
Immediate skin pigmentation	10000	340
Delayed skin pigmentation	30 - 45	298
Photoconjunctivitis	5	260
Photoceratitis	10	288

Based on the known threshold doses required to generate a detectable biological response, maximum UV dose levels for an eight hours exposure were defined. One standard agreed internationally is that of the American Conference of Governmental Industrial Hygienists (ACGIH). Approved exposure limits for 8 hours exposition of skin and eyes are given in Figure 1.

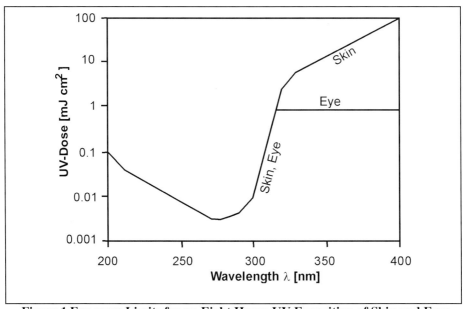

Figure 1 Exposure Limits for an Eight Hours UV Exposition of Skin and Eyes

In Figure 1 the UV dose is shown as a function of wavelength. The dose values strictly apply for monochromatic radiation. They are considered very safe and are considerably lower than those which can be experienced from sunlight exposure.
Dividing the minimum exposure dose limit (at 270 nm) by the exposure time will yield a limiting value of UV irradiance.

Table 2 UV Irradiance Limits at Different Exposure Times

Duration of Exposure per Day	Irradiance Limit mWcm^{-2}
8 hours	0.0001
1 hour	0.0008
10 minutes	0.005
1 minute	0.05
10 seconds	0.3
1 second	3

The irradiance limits summarised in Table 2 apply for unprotected skin and eyes. When working on UV curing units eye preservers and suitable clothes should be worn, in particular, when taking samples from machines while the UV lamps are operating.

(b) Shielding for UV equipment

The UV equipment manufacturers have established a number of simple design criteria in order to eliminate UV exposure risks [1,2,3].

These are usually based on primary or secondary shielding of the light source. The effectiveness of shielding is often tested by monitoring the intensity of the emitted visible light, which is proportional to the UV output.

Adequate primary shielding for UV light is provided by following measures:
- obstruct direct radiation from lamps or reflectors,
- dissipate UV light by multiple reflections in a long and narrow entry/exit tunnel,
- allow for maximum light absorption on the tunnel surface.

In practice that means:
- the light shield entry and the exit height should be kept as small as possible,
- a minimun tunnel length of 8 x entry slit height is required,
- where light reflection can occur, the inner tunnel surface should be matt black,

- when flexible substrates are used, the entry and exit should be taken through a suitable angle,
- curtain outfeed is recommended.

Figure 2 shows some typical tunnel designs.

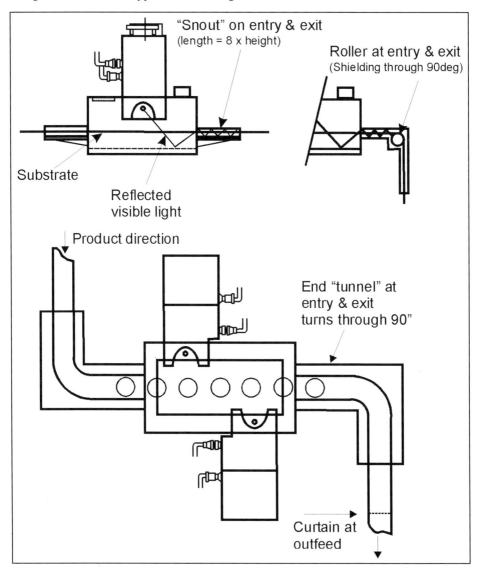

Figure 2 UV Shielding for Two- and Three-Dimensional Products

Where large three-dimensional objects are UV cured, primary shielding has to be provided by the irradiation chamber. A tunnel labyrinth or automatically operated doors are preferred technical solutions.

f parts of the primary shielding can be removed, interlock switches are fitted to the shielding so that lamps are turned off when the shields are opened.

In most cases primary shielding is adequate. However, secondary shielding can be required, e.g. when small radiation passes exist. Secondary shields, e.g. made from UV absorbing but transparent material, can be placed in strategic positions between the UV irradiator and any personnel in the vicinity.

(c) Shielding for heat

About 45% of the electric input power of a medium pressure mercury lamp is converted into heat. Heat is generated by the infrared part of the lamp spectrum and by the lamp jacket. The latter is called convected heat. In air-cooled systems most of the convected heat is dissipated within the air flow. Chapter VII gave a detailed description of heat management in UV curing processors. Water-cooled reflectors, lamp housings, shutters and aluminum plate heat sinks placed below the lamp and the substrate are commonly used to avoid overheating of the substrate and UV processor parts, thus preventing skin burning when touching them.

(d) Protection against ozone

Ozone (O_3) is a gas formed by UV-induced dissociation of oxygen. Ozone formation takes place when photons with wavelengths below 240 nm interact with molecular oxygen forming oxygen atoms. Oxygen atoms undergo rapid reaction with molecular oxygen. In the presence of a neutral collision partner M (N2, O2) ozone is generated:

$$O_2 + h\nu \rightarrow O + O \quad \text{at } \lambda < 240 \text{ nm} \quad (3)$$
$$O + O2 + M \rightarrow O_3 + M . \quad (4)$$

Ozone is a labile molecule which decomposes under normal pressure and room temperature with a time constant of about 20 min. Inhalation of the highly oxidising ozone causes headaches, fatigue, dryness of the throat and respiratory infections. The internationally agreed acceptable maximum ozone concentration level is 0.1 ppm by weight for an eight hour exposure. Figure 3 illustrates different biological effects of ozone as a function of the exposure time.

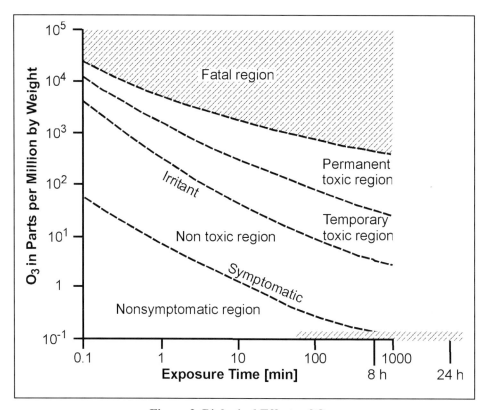

Figure 3 Biological Effects of Ozone

Ozone decomposition can be considerably accelerated by catalysts such as charcoal. The latter is sometimes used as an ozone filter in UV curing units. The most popular method to remove oxygen from the workplace is exhausting to the external atmosphere.

It should be pointed out that nitrogen inerting or applying an 308 nm excimer lamp as UV curing source are also measures to avoid ozone emission. Ozone-free quartz would be another solution but it is rarely used because absorption of the short wave part of the medium pressure mercury lamp reduces the cure rate.

(e) High voltage protection

Most of medium pressure mercury arc lamps operate at voltages well above the mains level. However, the risk of exposure to high voltages is reduced by using approved standards for construction and installation of the electric part of a UV processor.

In the case of electrode-less microwave-powered lamps a microwave detector is always supplied to detect any possible microwave leakage. The electrode-less system is shut down after 16 seconds when a microwave irradiance of 5 $mWcm^{-2}$ is detected.

2. EB Equipment Health and Safety

(a) X-ray formation by fast electron

The energy loss of fast electrons in matter is caused by particle collisions and radiation. Electron beams with energies between 30 and 300 keV, as generated by low-energy electron accelerators, are high-energetic enough to ionise matter. Their energy losses can be described by the concept of the stopping power.

The total mass stopping power $(S/\rho)_{tot}$ is defined as the total energy loss dE by collision and radiation for a pathlength dl in matter of the density ρ. For the electron energy range defined above, the total mass stopping power consists of two terms, the mass collision stopping power $(S/\rho)_{col}$ and the mass radiative stopping power $(S/\rho)_{rad}$:

$$(1/\rho)dE/dl_{tot} = (S/\rho)_{col} + (S/\rho)_{rad}. \qquad (5)$$

The first component of Eqn.5 includes all energy losses in particle collisions which directly produce secondary electrons and atomic or molecular excitation. The second component describes all energy losses of the primary electron which lead to bremsstrahlung (or X-ray) production.

Figure 4 schematically shows possible x-ray sources in electron beam curing. Some accelerated electrons are stopped in the copper support window. Others pass the holes in the support flange and penetrate the titanium foil, pass the nitrogen atmosphere in the gap between foil and coating, are partially absorbed in the coating and finally stopped in an aluminum backing.

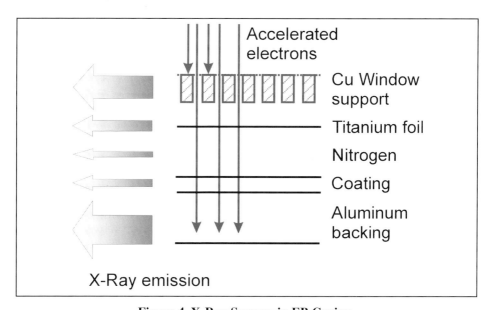

Figure 4 X-Ray Sources in EB Curing

As indicated in Table 3, for energies from 30 to 300 keV the radiative stopping power contributes between 0.1 and 1% to the total mass stopping power, i.e., between 0.1 and 1% of the total electron beam power is converted into x-ray power.

If a sequence of absorbers is assumed as given in Figure 4, 30 keV electrons are completely stopped in the titanium foil. This foil forms an x-ray source with a broad energy distribution below 30 keV and an x-ray dose rate distribution peaking in forward direction.

180 keV electrons penetrate the foil, the nitrogen and the coating. Finally, they will be absorbed in the thick, water-cooled aluminum backing. Now the window support flange, the aluminum absorber and the foil are the main x-ray sources. The energy of the x-rays emitted shows a wide distribution with a maximum below 180 keV. The x-ray dose rate is peaking in forward direction.

Table 3 Interaction of Electrons with Matter: Mass Collision and Stopping Powers [4]

Electron energy (keV)	Mass collision stopping power (MeVcm^{2}g^{-1})				Mass radiation stopping power (MeVcm^{2}g^{-1})			
	Ti	N$_2$	PMMA*	Al	Ti	N$_2$	PMMA*	Al
30	6.50	8.56	9.40	7.29	1.15×10^{-2}	3.77×10^{-3}	3.39×10^{-3}	7.06×10^{-3}
100	2.87	3.66	4.01	3.18	1.27×10^{-2}	4.01×10^{-3}	3.62×10^{-3}	7.48×10^{-3}
200	1.97	2.49	2.72	2.17	1.42×10^{-2}	4.55×10^{-3}	4.13×10^{-3}	8.34×10^{-3}
300	1.67	2.10	2.29	1.84	1.60×10^{-2}	5.23×10^{-3}	4.75×10^{-3}	9.49×10^{-3}

* Polymethylmethacrylate

Another interesting phenomenon is the scattering of secondary electrons in the nitrogen gas below the exit foil.

Emission from excited nitrogen molecules indicates the presence of electrons. Due to collisional scattering an electron cloud is formed around the exit window. This electron cloud does not only show the spatial distribution of secondary electrons but is also a source of x-ray emission. Figure 5 shows the visible part of the nitrogen emission, which was obtained by irradiation of air with 200 kev electrons.

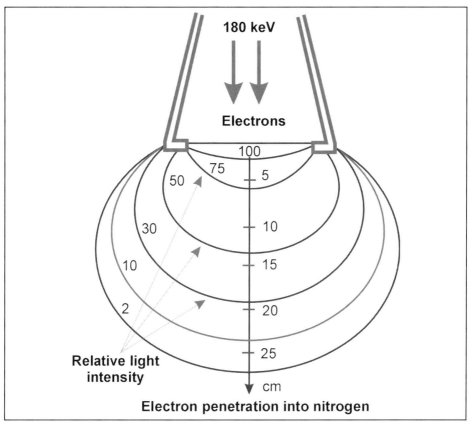

Figure 5 Visible Light Generated by Electrons in Air [5]

(b) Biological action of x-rays

Absorption of x-rays exhibiting energies below 1 MeV leads to electron formation via photoeffect and Compton scattering. The electrons formed after x-ray absorption lose their energy by Coulomb interaction with the electrons of the surrounding atoms and molecules and are finally thermalized. Thereby, ionisation and excitation takes place in a limited spatial volume. A very non-homogeneous, cluster-like structure of ionisation and excitation centres is formed along the path of the secondary electron and preferably at the end of its slowing down range. There is a large body of radiobiological data to show that such a track structure is of great significance for the biological action of ionising radiation.

One important result of electron absorption in cells is DNA damage, either initiated by radiolysis products (radicals, solvated electrons) in the aqueous phase (indirect effect) or by direct ionisation and deterioration of the DNA. In most cases, radiation damage of DNA is efficiently and rapidly repaired but there is also some probability that permanent damage can occur. It is the permanent cell damage that can lead to cancer formation or genetic mutations.

On a macroscopic scale, a certain radiobiological effect is often described as a function of the absorbed dose. Some radiation damage effects such as skin burning and the general radiation syndrome seem to be caused after a certain threshold dose has been reached. On the other hand, cancer formation seems to be a stochastic process occurring without a threshold dose. Therefore, radiation exposure should be kept as low as reasonably achievable [6]. For human radiation exposition the International Commission on Radiological Protection recommends a dose threshold value of 1mSv/year. This value is about 30% higher than that obtained by exposure to the natural cosmic and terrestrial radiation in the main part of Germany.

(c) Radiation shielding of low-energy electron accelerators

Radiation shielding of low-energy electron accelerators used in curing should be done in a way to ensure that:

a dose rate of 1 µSv/ h is not exceeded in a 5 cm distance from all operator-accessible surface parts of the electron processor shielding.

The dose rate obtained outside the shielded radiation processor is determined by the efficiency of the x-ray absorption of the shielding material, the geometrical distance from the x-ray source and the leakage of the shielding construction.

For x-rays with energies below 300 keV the photoeffect is the main absorption mechanism. Its absorption cross-section depends on the x-ray energy and rises with the atomic number. Therefore, lead laminated on steel is the shielding material preferred for low-energy electron processors.

The dose rate attenuation I/I_o caused by an absorber of the thickness x and the mass absorption coefficient μ is given by $I/I_o = \exp(-\mu x)$. Figure 6 shows the mass absorption coefficient μ for lead as a function of the x-ray energy. It is strongly energy dependent and also shows typical inner shell absorption effects (K and L shell).

As an example, 10 mm lead would decrease the dose rate of 100 keV x-rays by a factor of 1.4×10^{-5}.

However, to calculate technical shieldings for x-ray sources such as electron processors is more complex. In such cases, the use of suitable tables is very useful [7].

There are some rough practical guidelines for lead shielding of low energy electron accelerators: 150 keV electrons require between 8 and 12 mm, 250 keV between 25 and 35 mm lead shielding. In particular, this is the case for non-scattered x-rays. In processor areas where only scattered x-rays hit the shielding, the lead thickness can be reduced.

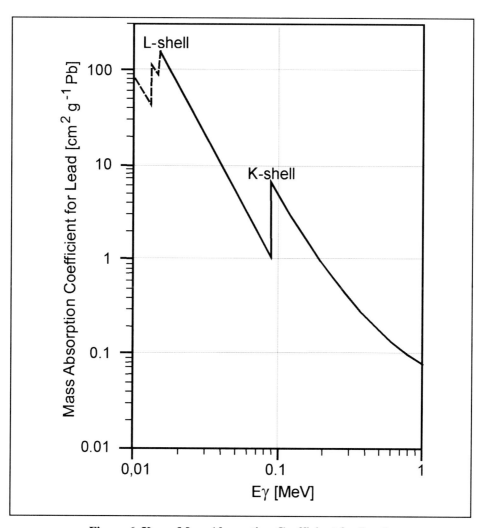

Figure 6 X-ray Mass Absorption Coefficient for Lead

Additionally, a geometric effect leads to a decrease in the x-ray dose rate. The dose rate falls off with the distance d from the source according to $I/I_o = 1/d^2$.

Most electron processors used in curing are "self-shielded", i.e. the electron (and x-ray) source is completely enclosed by shielding. Removable parts of the shielding have to be equipped with safety interlocks so that the high-voltage of the accelerator is turned off when the shielding is opened.

In standard EB curing applications, not only the radiation source but also the conveyor system, or at least a part of it, have to be shielded as well. For flat rigid substrates such as board or plates different technical solutions are now in use. Some of them are shown in Figures 7 and 8 [8].

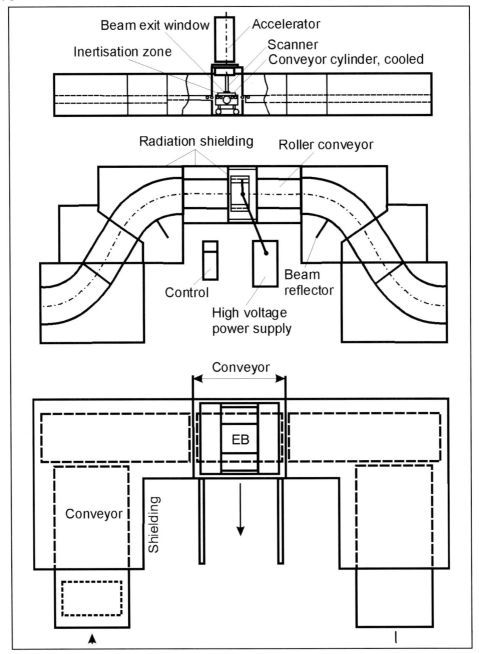

Figure 7 X-ray Shielding Through an Ω- Loop or Angle Transfer

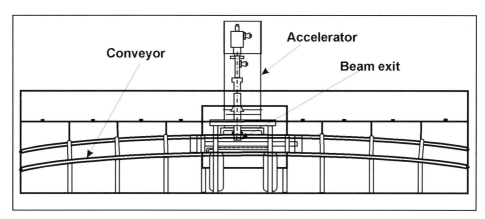

Figure 8 X-ray shielding Trough Bent Conveyors

Shielding is much simpler when flexible substrates are irradiated. Figure 11 in Chapter VI shows a possible technical solution [9]. The web enters horizontally from the left side and comes vertically down through the electron beam. The web exit is again horizontal. Bending of the web material must not necessarily amount to 90°. If the entry/exit slit width can be kept in the range of 1 to 2 mm, a smaller bending angle can also be used to achieve suitable shielding.

Another elegant way to provide perfect x-ray shielding for web-fed curing is a metal drum placed below the electron beam exit [5]. The drum is lead shielded and water-cooled. It not only absorbs nearly all non-scattered x-rays but also serves as substrate feed. The entry/exit channel of the substrate can be kept long and narrow so that multiple scattering and effective absorption of x-rays takes place.

REFERENCES

1. *E. Horwood, Proceedings Radcure Europe `87, p. 2-1 (1987)*
2. *R. Denney, Proceedings RadTech Europe`95, p. 363 (1995)*
3. *A. Beying, IST-UV-Seminar Münster (1998)*
4. *Radiation Dosimetry: Electron Beams with Energies between 1 and 50 MeV, ICRU Report 35, Bethesda 1984*
5. *P. Holl, E. Föll, Proceedings RadTech Europe`89, p. 609 (1989)*
6. *H. Kiefer, W. Koelzer: Strahlen und Strahlenschutz, Springer-Verlag, Berlin, Hieidelberg, 1987*
7. *C. Dimetrijewic: Praktische Berechnung der Abschirmung radioaktiver und Röntgenstrahlen, Verlag Chemie, Weinheim, 1972*
8. *P. Holl, E. Föll, Proceedings RadTech Europe`91, p.478 (1991)*
9. *P.M. Fletcher, Proceedings RadTech Europe`93, p. 637 (1993)*

INDEX

A

absorbed dose	108-112, 114
abstractable hydrogen	6
accessible electron energy	142
Acrylate & methacrylate systems	1-7
actinometry	121, 124, 126-130
alkaline earth oxides	54
amine synergists	6
Applications of thin film dosimeters	117
argon	54, 65, 70
arylsulphonium salts	19, 21

B

"back scattering"	110, 111
bisphenol A	7
Brönstedt acids	7, 19, 24

C

calibration of dosimeters	115, 116, 117
capacitors	67
cationic curing initiation	19
- arylsuphonium salts	19
- Bronstedt acids	7, 19, 24
- Lewis acids	7, 19, 24
- protonation of oxirane ring	20
- sensitisers	21
- solvent cage	19
- triarylsulphonium salts	19
cationic polymerisation	7, 24, 83
- bisphenol A	7
- epoxides	7, 20
- epoxy-functional silicone polyethers	8
- polyols	3, 7, 200, 206, 225
- post cure	8, 67, 196, 205, 206, 208, 233
- vinyl ethers	7, 21
cationic systems	7-10, 19-21
chain propagation	18
chromophores	10, 14, 19

C (contd)

coatings on rigid substrates	145
coinitiators	6
cold excimer radiation	59
cold or thermionic cathode lamps	50-53
collimeter	129
colorimetry	108
cure characterisation	243-263
cure speed	215-236
cure yield	117
curing applications (EB)	135

D

Degree of cure & cure speed	195, 215-236
- cure speed	215-236
- "dark reactions"	196, 205, 206, 208, 214, 233
- decay of radical concentration	207
- degree of cure characterisation	243-262
- determination of propagation & termination rate constants	236
- DMA & Rheometry	260-262
- effect of temperature	209-215, 225, 248
- factors affecting cure speed in cationic systems	235
- field cure tests	243-245
- fluorescence probe	254
- FTIR spectrometer	198
- Kinetic factors in EB curing	237-243
- kinetic perameters	196
- kinetic perameters in cationic curing	226, 237
- laboratory cure tests	245
- measurement	243
- molar heat	254
- multiple exposure	225
- nitrogen inerting	224
- oxygen	204, 222, 223, 224
- photo-calorimetry (DSC)	253
- photoinitiator IC 369	199
- physical factors	221
- pigmentation	221
- polymerisation at radical initiation (EB)	238-243
- polymerisation rate & irradiance	201
- reactive sites	195

D (contd)
- residual reactive sites & monomer & oligomer functionality 195
- spectral overlap 220
- spectroscopy 246-253, 256-259

delivered dose 141, 142, 153
deprotonation 22
depth dose distribution 13, 14
dielectric discharge 58, 83, 84, 85, 86
difusional motion 15
dimercations 22
Donor acceptor complexes 9, 10
- chromophores 10, 14, 19
- depth profile 14
- donor acceptor pairs 10
- electron deficient unsaturation 10
- excess electron charge 10
- vinyl ethers 7, 21

Dosimetry for EB & UV curing
- dose rate & uniformity 114, 115, 117
- photons 14, 100, 114
- radiant energy 55, 56, 76, 78, 79, 86, 101-107, 117, 118
- radiation intensity 107

dosimetry low energy electron beam 108
- "absorbed dose" 108, 112, 114
- applications of thin film dosimeters 117
- back scattering 110, 111
- calibration 115, 116, 117
- calorimetry 108
- cure yield 117
- depth dose distribution 110, 112, 118, 139
- dosimeter response 112, 114
- dosimetry with thin film dosimeters 112
- gel scanner 114
- interactions of electrons matter & dose 108
- linear stopping power 109
- mass collision stopping 110
- mass stopping power 109
- micro densitometer 115
- optical absorption thin film dosimeters (EB) 113, 127
- paramagnetic response EPR 115

D (contd)

- quantities relevant to EB dosimetry	109
- Radiochromic film dosimeters	114, 131
- Suppliers of film dosimeters	113
- surface area rate	117
dosimetry UV radiation	119
- actinometry	124, 126, 127, 128, 129, 130
- dose (UV dose)	120
- photochemical conversion	126
- photon detectors	122
- photon energy	121
- photonic qualities	120
- quantities & units related to radiation dosimetry	121
- radiant energy	55, 56, 76, 78 79, 99, 101-107
- radiometers	122, 123, 124
- radiometric quantities	119, 128
- thermal detectors	122
- uv sensitive films & labels	130

E

effect of temperature on cure	209, 225, 261
electrical characteristics - medium pressure lamps	55, 56
electron avalanche	98
Electron beam curing - Free radical polymerisation	22-24
- acrylate anions	22, 23
- Brönstedt acid	7, 19, 24
- Cationic polymerisation	24
- dimercations	22
- iodonium	24
- radical cations	22, 23
- resonance stabilised dimer anions	22
- salt cation	24
- solvated electrons	22
- sulphonium	24
Electron Beam Curing Equipment	135-158
"Ebogen" processor	153, 154
"Microbeam" accelerator	156, 157
- "spurs"	135
- accessible electron energy	142
- coatings on rigid substrates	145
- curing applications	135

E (contd)

- delivered dose	141, 142, 153
- EB curing characteristics	138
- Electrocurtain processor with selfshield assembly	150, 151
- electron acceleration	12, 13, 14, 135, 143
- electron penetration range	141
- electron processor	137, 150, 151, 152
- Electron processor units - Definitions & units	140
- linear cathode electron accelerators	149
- low energy electron accelerators	136, 137, 146, 147, 152, 153
- maximum beam power	143
- multifilament cathode electron accelerator	154, 155
- non porous substrates	145
- penetration depth	135
- rotational excitation	135
- types of industrial low energy electron processors	142
- vacuum diodes	152
- Wehnelt cylinder	148
electron deficient unsaturation	10
electrostatic (Coulomb) interaction	13
emission from microwave excited discharge	49, 50
epoxides	7, 20
epoxy functional silicone polyether	8
excess electron charge	10
excimer lamps	12, 53, 77, 79, 83, 101-104
excimer light sources	12, 53, 77, 79, 85, 86, 89, 91, 92, 93, 94, 95, 96, 98, 129
exposure limits for skin & eyes	266

F

field cure tests	243-245
fluorescence probe	255
foil	54, 59
Free radical cure	15, 17, 18
- acrylate/methacrylate systems	2, 3, 4, 14, 15, 22, 23
- chain propagation	18
- diffusional motion	15
- fluorescence & quenching by oxygen	15, 60, 83, 204, 222, 223, 252
- intersystem crossing	15

F (contd)

- maleate/vinyl ether systems	8, 15
- Norrish Type I & II reactions	15, 16, 17
- photon absorption	2, 3, 4, 14, 15, 43, 107, 121, 122
- radical formation from photoinitiators	15, 18
- Solvent case	15, 19
- termination	17, 18
FTIR spectrometers	198

G

Gallium doping	49, 55
gas halide excimers	12, 89, 91, 92
gel scanner	114
glow & arc discharge	44-48, 50, 51, 56, 76, 77, 79 80, 83, 101, 198
Grotion energy transition of mercury atoms	45, 46

H

halides	49, 50, 55, 73, 83, 89, 92, 98, 100
halogens	70, 83, 85, 90, 92
Health & Safety	265-277
- biological effects of ozone	270
- exposure limits for skin & eyes	266
- high voltage protection	270
- photobiological effect of UV radiation	266
- protection against ozone	266
- radiation shielding of low energy electron accelerators	274
- shielding for heat	269
- shielding for UV equipment	267-268
- x-ray	271, 273, 275, 290, 291
Hybrid curing	21
- acrylates	2, 3, 4, 15, 21, 22, 23
- arylsulphonium	19, 21
- interpenetrating networks	21
- vinyl ethers	7, 21

I

impedance load	50, 65
inductive & capacitive load	45, 65
Industrial Applications of Radiation Curing	32-42
Industrial low energy electron processors	142

I (contd)
Initiation of curing - photons & electrons
- depth dose distribution — 13, 14
- electrons — 12, 13, 14, 135, 143
- electrostatic (Coulomb) interaction — 13
- excimer lamps — 12, 53, 77, 79, 83
- gas halide exciplexes (excimers) — 12, 89
- low pressure medium lamps — 12, 50-82, 83, 85, 86, 88, 89, 91
- mercury discharge — 12
- nitrogen laser — 83
- photons — 2, 3, 4, 12, 14, 15, 22, 23, 107, 121, 122
- power densities — 12

interactions of electrons — 108
interpenetrating networks — 21
intersystem crossing — 15
iodonium — 14
IR emission & absorbance — 43, 58
iron doping — 49, 55
irradiance pattern — 61, 62, 64, 69, 74, 75, 77, 120, 231
irradiator — 71, 74, 76

K
kinetic factors in EB curing — 237-243
kinetic perameters — 189

L
laboratory cure tests — 245
lamp characteristics & makers — 52
lamp life — 70
Lewis acids — 7, 19
light source geometry — 44
linear cathode electron accelerators — 149
linear lamps — 60
linear stopping power — 109
low energy electron accelerators — 136, 137, 146, 147, 152, 153
low energy electronbeam dosimetry — 108
low pressure medium lamps — 12, 50-82, 85, 86, 88, 89, 91, 93

M
Maleate/vinyl ether systems — 8, 15
mass stopping power — 109, 110

M (contd)

maximum beam power	143
Medium pressure mercury arc lamp	53-81
mercury discharge	12, 45, 46
mercury vapour	44, 45, 46, 50, 65, 86, 92
Microbeam accelerator	156, 157
microdensitomer	115
microwave discharges	71, 72, 74, 75, 76, 93, 96 97, 98, 100
molar heat	254
monochromatic sources	84
Monochromic lamps	83-104
multifilament cathode electron accelerator	154, 155
multiple exposure	225

N

nitrogen inerting	224
nitrogen laser	83
non parous substrates	145
Norrish Type I & II reactions	15, 16, 17

O

optical absorption thin film dosimetry	113, 127
optical fibres	60
oxirane ring	20
oxygen inhibition	15, 60, 83, 204, 222, 223, 224
ozone	69, 127

P

paramagnetic response EPR	115
penetration depth (EB)	135
photobiological effects of UV radiation	266
photocalorimetry (DSC)	253
photochemical conversion	126
photoinitiators IC 369	199
photon2, 3, 4, 12, 14, 15, 22, 23,	44, 107, 121, 122
photonic qualities	120
physical factors in cure	221
pigmentation	221

P (contd)
Polychromatic Light Sources for UV Curing   50-82
  1. Low pressure mercury lamp   12, 50-82, 83, 85, 86
                                                   88, 89, 91
       - cold or thermionic cathode lamps   50
       - excimer light sources   12, 53, 77, 79-85, 86, 89
                                                   91-96, 101, 103
       - impedance load   50
       - lamp characteristics & makers   52
       - quartz   50, 51, 94
       - resonance radiation   51, 83, 88
       - spectral output   44, 48, 52, 55, 57, 220
  2. Medium pressure mercury arc lamp   53-81
       - "dark period"   67, 196, 205, 206, 208
                                            214, 233
       - alkaline earth oxides   54
       - argon   54, 65, 70
       - capacitors   67
       - cold excimer radiation   59
       - dielectric discharges   58, 83, 84, 93
       - electrical characteristics & definitions   55, 56
       - electrodeless lamp   73, 74, 76
       - foil   54, 59
       - halogens   70, 83, 95, 90, 97
       - impedance   50, 65
       - irradiance pattern   61, 62, 64, 69, 74, 75, 77
                                          120, 231
       - irradiator   71, 74, 76
       - lamp life   70
       - linear lamps   60
       - microwave discharge   71, 72, 74, 76, 83, 93, 96
                                          97, 98, 100
       - ozone free lamps   69
       - power supply & lamp control   65, 66, 69, 76
       - radiant energy   55, 56, 76, 78, 79, 99, 101, 107
       - reflectors & lamp cooling   57-63
       - spectral output   44, 55, 57, 58, 65, 75, 220
       - thoriated tungsten rods   54
       - transducers & thyristors   67, 68
       - UV sensor   68
       - vitreons silica   53, 54, 70, 72
       - xenon   77, 79, 80, 87-89, 99

P (contd)
    3. Monochromatic Lamps   83-104
        UV & VS emission from excimers   84, 86, 88, 94, 99, 102
        - collimator   129
        - dielectric barrier discharge   58, 83, 84
        - excimer lamps v medium pressure   101-104
          mercury lamps
        - excimers & light   12, 53, 77, 79, 85, 86, 89
                                                                91-96, 98, 129
        - monochromatic sources   84
        - specification for excimer lamps   101
Polychromatic UV Lamps   43-82
    - electron avalanche   98
    - electrostatic (Coulomb) forcer   45
    - emission from microwave excited discharge   49, 50
    - Gallium doping   49, 55
    - glow & arc discharge   44-48, 50, 51, 56, 76, 77
                                                             79, 80, 83, 101, 198
    - Grotian energy transition of mercury atoms   45, 46
    - halides   49, 50, 55, 73, 83, 89, 92
                                                                                98, 100
    - inductive & capacitive load   45, 66
    - IR emission & absorbance   44, 58
    - iron doping   49, 55
    - light source geometry   44
    - mercury gas discharge   12, 45, 46
    - mercury vapour   44, 46, 50, 65, 86, 92
    - micro discharges   86
    - radiant energy   55, 56, 76, 78, 79, 99, 101, 107
    - radiative transitions   47
    - recombination   45, 48
    - self absorption   45
    - spectral output & modification   44, 48, 52
    - spin quantum   47
    - spot curing   80, 81
    - thermionic electron emission   45
    - three dimensional application   80
    - voltage - current characteristics of a   44
        gas discharge
polyether acrylates   6
polymerisation rate & irradiance   201
polymerisation rate at radical initiation (EB)   238-247

P (contd)
Polyols	3, 7, 200-206, 225
- bifunctional & polyfunctional acrylates	4
- chain transfer agents	7
- HDDA	
- PETA	
- TMPTA	
- TPGDA	
powder coatings	9
power densities	12
propagation & termination rate constants	208
protection against ozone	266, 270

Q
quantities relevant to EB dosimetry	109, 121
quartz	50, 51, 94

R
radiant energy	55, 56, 76, 78, 79, 99, 101-107
	107, 118
Radiation curing growth	27, 28
Radiation curing light sources	49, 107, 118
radiation intensity	107
radiation shielding of low energy electron accelerators	274
radical cations	22, 23
radical formation from photoinitiators	15, 18
radiochromic film dosimeters	117
radiometers	122, 123, 124
radiometric quantities	119, 128
reactive sites	195
reactive thinners	3
reactive transitions	47
recombination	45, 48
reflectors & lamp cooling	57-63
resonance radiation	51, 83, 88
resonance stabilised dimer anions	22, 23
rotational excitation	135

S
"spurs" (EB)	135
salt cation	24

S (contd)

self absorption	45
sensitisers	21
shielding for heat	269
shielding for UV equipment	267-268
solvated electrons	22
solvent cage	15, 19
specification for excimer lamp	101
spectral output & overlap	44, 48, 52, 55, 57, 65, 75, 234
spectroscopy	246-253, 256-259
spin quantum	47
spot curing	80
sulphonium	24
suppliers of dosimeters	113
surface area rate	117

T

termination	17, 18
thermal detectors	122
thermionic electron emission	45
Thiolene systems	11
- olefinics	11
- polythiols	11
thorated tungsted rods	54
three dimensional applications	80
transducers & thyristors	67-68
triaryl sulphonium salts	19

U

unsaturated polyesters	11
- glycols	11
- organic diacids	11
- phthallic anhydride	11
- styrene/polyester	11
UV & VS emission from excimers	84, 86, 88, 89, 94, 97
UV light sources	49
UV radiation dosimetry	119
UV sensitive films & labels	130
UV sensor	68

V

vacuum diodes	152

V (contd)
vinyl ethers 7, 21
vitreous silica 53, 54, 70, 72
voltage current characteristics of gas discharge 44

W
water soluble monomers
    - epoxy acrylates 6
    - polyester acrylates 6
    - urethane acrylates 6
Wehnelt cylinder 148

X
x-ray 271-273, 275-277